Making Semiconductors

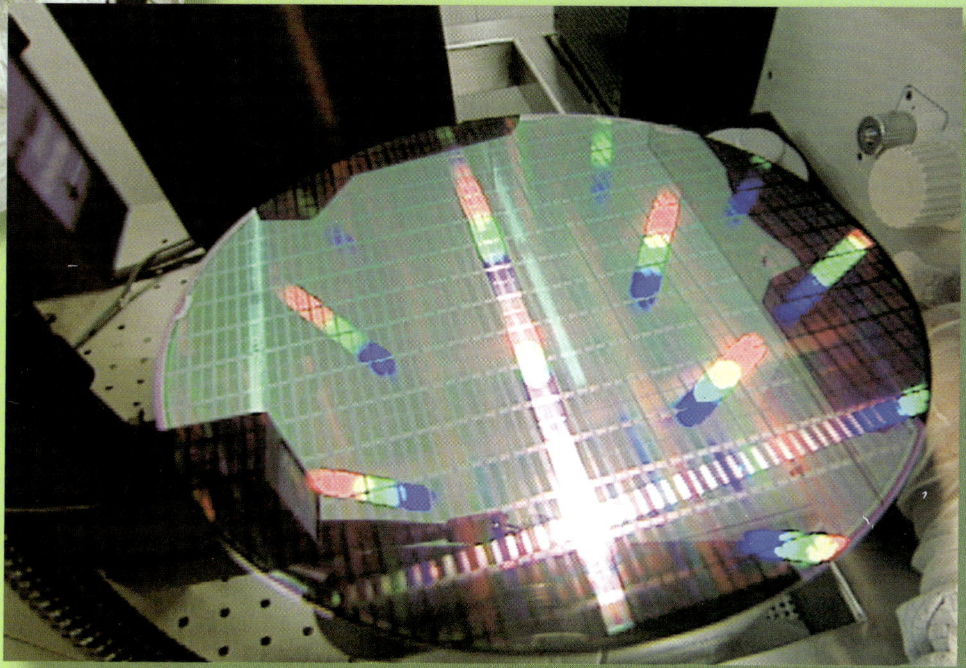

SCIENCE in a technical world

A Project of the Education Division of the American Chemical Society

W. H. Freeman and Company
New York

Science in a Technical World Project Team
Co-Principal Investigators: Sylvia Ware, David Lavallee, Kenneth Chapman
Project Director: Ann Benbow
Production Director: Colin Mably
American Chemical Society: Sylvia Ware, Ann Benbow, Janet Boese, Gary Dikeos, Helen Herlocker, Guy Belleman, Michael Tinnesand
Writers: Jack Ballinger, Ann Benbow, Helen Herlocker, Wayne Morgan, Carol Muscara, Marti Tomko, Emmett Wright, Robert Bernoff, Ken Weese
Editors: Marcia Vogel, Ann Benbow, Colin Mably
Project Evaluator: Robert Bernoff
Assessment Consultant: Gail Goldberg
Video Coordinator: Colin Mably
Video Production: Take One Video
CD-ROM Development: CTS, Inc., Sonalysts, Inc., Take One Digital Multimedia
Safety Reviewer: W. H. Breazeale
Content Advisor: Mark A. Holcomb

National Science Foundation Visiting Committee
Carlo Parravano, Chair; Barbara Wrobleski, Michael Farmer, Mickey Sarquis, Nancy Dillon, Robert Maleski, Felicia Graves, Toby Horn

W. H. Freeman and Company Book Team
Publisher: Michelle Russel Julet
Sales and Marketing Consultant: Arthur C. Germano
Development Editor: Rebecca Strehlow
Project Editor: Mary Louise Byrd
Text Designer: Blake Logan
Cover Designer: Patricia McDermond/Blake Logan
Cover Illustration: Roy Weiman
Illustration Coordinator: Lou Capaldo
Illustrations: Proof Positive/Farrowlyne Associates, Inc.
Photo Research: Proof Positive/Farrowlyne Associates, Inc.
Production Coordinator: Julia DeRosa
Production and Composition: Proof Positive/Farrowlyne Associates, Inc.
Manufacturing: RR Donnelley & Sons Company

Library of Congress Cataloging-in-Publication Data

Making semiconductors / a project of the Education Division of the American Chemical Society.
p. cm. — (Science in a technical world)
ISBN: 0-7167-4036-2
1. Semiconductors—Design and construction. 2. Integrated circuits—Design and construction. I. American Chemical Society. Education Division. II. Series.

T65 .A84 2001
621.3815'2—dc21 2001033575

Development of the *Science in a Technical World* program was supported, in part, by the National Science Foundation under Grant Nos. DUE 9454564 and DUE 9752102. Opinions expressed are those of the authors and not necessarily those of the Foundation.

Printed in the United States of America
First printing 2001

Science in a Technical World Industry Contacts

ROBERT HOFSTADER (industry consultant); TERRY WEBB, Quality Assurance, Pepsi Bottling Group; JEFF FLYNN, Kleentek; RONALD BARBARO, Northern Virginia Community College; STEVEN CAWTHRON, Leesburg Wastewater Treatment Facility; KENNETH GLANZ, Appleton Papers, Inc.; MARIA SPINU, DuPont Research and Development; PATRICIA LISTER, DuPont Marshall Laboratories; PETER BICK, Eli Lilly and Company; MARK A. HOLCOMB, Dominion Semiconductors; PAMELA MCCARTHY, Civista Medical Laboratory; JENNIFER FERNS, Federal Bureau of Investigation; JERGEN VON BREDOW, Food and Drug Administration; FRED CONFORTH, Phillips 66 Company

Science in a Technical World Pilot Teachers

JEFF ALISAUCRAS, South Carroll High School, Sykesville, MD
JAMES BAUER, McKinley High School, Honolulu, HI
BRITTANY BEHRENS, Coronado High School, Scottsdale, AZ
BARBARA BILLINGS, Milton High School, Alpharetta, GA
BILL CALLAHAN, Watsonville High School, Watsonville, CA
JIM DIXON, Silver Lake Regional High School, Kingston, MA
KATHLEEN DOMBRINK, McCluer North High School, Florissant, MO
RONALD DUNAWAY, Monteray High School, Lubbock, TX
JONATHAN DUNSKI, North Carroll High School, Hampstead, MD
DOUG ENGLAND, Wallace High School, Wallace, ID
ROBERT FOUR-HOGUE, South Carroll High School, Sykesville, MD
JACK GOLDSTON, Ross High School, Fremont, OH
ALAN GONZALES, Lakewood High School, Lakewood, CO
KRISTA GRAY, McDowell Senior High School, Erie, PA
BERNIE HERMANSON, Harlan High School, Harlan, IA
NICK HOFFMAN, Wallace High School, Wallace, ID
JOCELYN HOLLIS, Chester High School, Chester, PA
SAMUEL HOLTZMAN, Pilgrim High School, Warwick, RI
JULIA JANOWICH, Francis Scott Key High School, Union Bridge, MD
CINDY JONES, Jefferson Senior High School, Edgewater, CO
LEAH LINDBLOM, Arvada West High School, Arvada, CO
MARY LINDIMORE, Arvada West High School, Arvada, CO
KAREN LUNIEWSKI, North Carroll High School, Hampstead, MD
DIANNE MCCLEARY, Montwood High School, El Paso, TX
TOM MCCOY, Keystone High School, Knox, PA
SUE MATTESON, McDowell Senior High School, Erie, PA
SHARON MUENCHOW, Arvada West High School, Arvada, CO
ELLIE MUNIZ, Ridgeview High School, Columbia, SC
MIKE MURRAY, Silver Lake Regional High School, Kingston, MA
MARK NALE, Tyrone Area High School, Tyrone, PA
KRISTEN NELSON-MCKENZIE, Thornridge High School, Dolton, IL
BEN NORTON, Bergen County Technical High School, Hackensack, NJ
SCOTT OLSON, Red Rocks Community College, Lakewood, CO
MARIA PACHECO, Buffalo State College, Buffalo, NY
ROBIN PARKINSON, Northridge High School, Ogden, UT
SHELLY PERETZ, Thornridge High School, Dolton, IL
RUSS RAPOSE, Pilgrim High School, Warwick, RI
CANDY RUZIECKI, Chester Senior High School, Chester, SC
BARBARA RYDEN, McDowell Senior High School, Erie, PA
THOMAS SANDHAM, North Kingston High School, North Kingston, RI
WAYNE SILL, Hutchinson High School, Hutchinson, KS
GEOFF SMITH, Watsonville High School, Watsonville, CA
BILL SPEED, Need School, Kailua, HI
TONY TAYLOR, Jefferson County Open School, Lakewood, CO
KATHLEEN THOMPSON, Anthony Jr/Sr High School, Anthony, TX
SCOTT THOMPSON, Arvada High School, Arvada, CO
LILLIE TUCKER-AKIN, Tupelo Christian School, Tupelo, MS
BARBARA WALKER, Ottumwa Community Schools, Ottumwa, IA
LARRY WEATHERWAX, Mclaughlin School, Alaska
KEN WEESE, Montwood High School, El Paso, TX
GARY WHITLING, Keystone High School, Knox, PA
HAROLD WILSON, Medomak Valley High School, Waldoboro, ME
DELORES WRIGHT, Chester High School, Chester, PA
DOUG YUST, Francis Scott Key High School, Union Bridge, MD

Teacher Review Panel

Bernie Hermanson
Nick Hoffman
Mark Nale
Candy Ruziecki
Barbara Walker
Larry Weatherwax
Ken Weese

contents

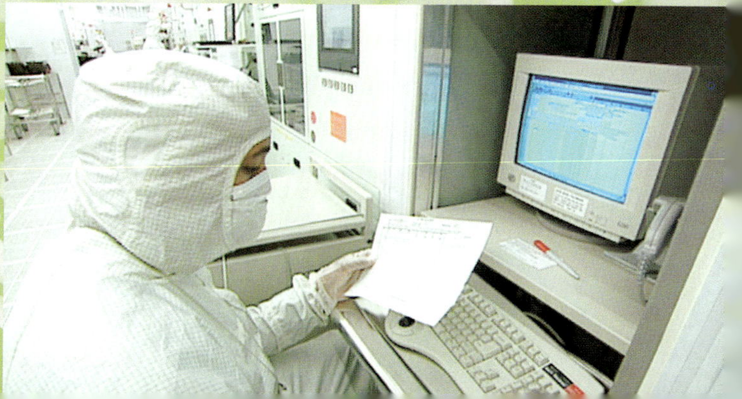

Multimedia Resources

Making Semiconductors Video: The 14-minute video gives you an overview of the silicon chip fabricating process within a major semiconductor corporation from raw materials through the many stages of production.

Making Semiconductors: The CD-ROM contains an introduction to the importance of semiconductors in today's world, an overview called Making An Integrated Circuit, a Technician Tasks section explaining the procedures and tests technicians perform at key points in the process, a Practice section that allows you to learn semiconductor protocols and procedures, an Assessment section to help you demonstrate your semiconductor making knowledge and skills, and a Reference section containing key terms and definitions related to the semiconductor making process.

Semiconductors on the Internet: A list of Web sites related to the semiconductor industry, which you will find useful throughout your work on this module, is included at the back of this book.

www.whfreeman.com/STW Additional resources for instructors and students are available online at the *Science in a Technical World* support Web site.

Science Connections Science Connections boxes offer background information and explanations to enhance your understanding of important science concepts covered in this module. You will find them near the places in the book where a key science concept is first introduced.

important notice

Science in a Technical World modules are intended for use by students in the classroom laboratory under the direct supervision of a qualified science teacher. The experiments described in this book involve substances that may be harmful if they are misused or if the procedures described are not followed. Read cautions carefully and follow all directions. Do not use or combine any substances or materials not specifically called for in carrying out experiments. Other substances are mentioned for educational purposes only and should not be used by students unless the instructions specifically so indicate.

The materials, safety information, and procedures contained in this book are believed to be reliable. This information and these procedures should serve only as a starting point for good laboratory practices, and they do not purport to specify minimal legal standards or to represent the policy of the American Chemical Society. No warranty, guarantee, or representation is made by the American Chemical Society as to the accuracy or specificity of the information contained herein, and the American Chemical Society assumes no responsibility in connection therewith. The added safety information is intended to provide basic guidelines for safe practices. It cannot be assumed that all necessary warnings and precautionary measures are contained in the document and that other additional information and measures may not be required.

welcome!

What Are the Goals of *Science in a Technical World*?

Before you begin working with this module, it is important for you to understand what the entire *Science in a Technical World* program is set up to accomplish. Each *Science in a Technical World* module is designed to help you to

- gain an understanding of the **science technology** used in a variety of industries and to become skilled in using science technology to solve problems that actually occur in industry
- become knowledgeable about the **industrial process** in general and compare the employment opportunities that a number of science technology industries offer
- work as a team in a **virtual workplace** setting

What Is Science Technology?

Science technology may be a term that is unfamiliar to you. It refers to the applied science that is used in a wide variety of settings, industry being one.

For example, science technology includes the laboratory equipment and instruments used to monitor industrial processes. These can be the same kind of "wet lab" equipment you may have used in other high school science courses (such as chemistry, biology, or the earth sciences), or they may be very sophisticated, computerized instruments.

This program will give you the chance to learn science technology in an industrial context by letting you and your classmates solve actual problems from different industries.

What Is the Industrial Process?

Industry is the process by which products are made. Some industries with which you may be familiar include the automotive industry (manufactures cars), the food industry (processes food products), the pharmaceutical industry (makes medications), the paper industry, and a number of others.

Science in a Technical World problems are taken from some of these familiar industries, as well as from others that you may not know as much about. The industries selected for the entire program were chosen to allow you to gain science technology skills and knowledge in the biological, chemical, physical, and earth sciences.

What Is the Virtual Workplace?

The **virtual workplace** is a term used in the *Science in a Technical World* program to describe a new way of thinking about your classroom, your teacher, and your classmates. As much as possible, the work that you do throughout the program will mirror what actually happens in industry. This means you will work as teams to solve problems and will be dependent on the work of other teams, at times, to achieve results. You will do research to find information from a variety of sources (including a CD-ROM, a video, and the Internet) to help you solve problems. You will work according to a timeline that all of you establish. You will also keep very careful records of your work, which you will need to have witnessed and signed off on a regular basis. Your teacher's role will shift from an instructor to more of a manager or supervisor of your work.

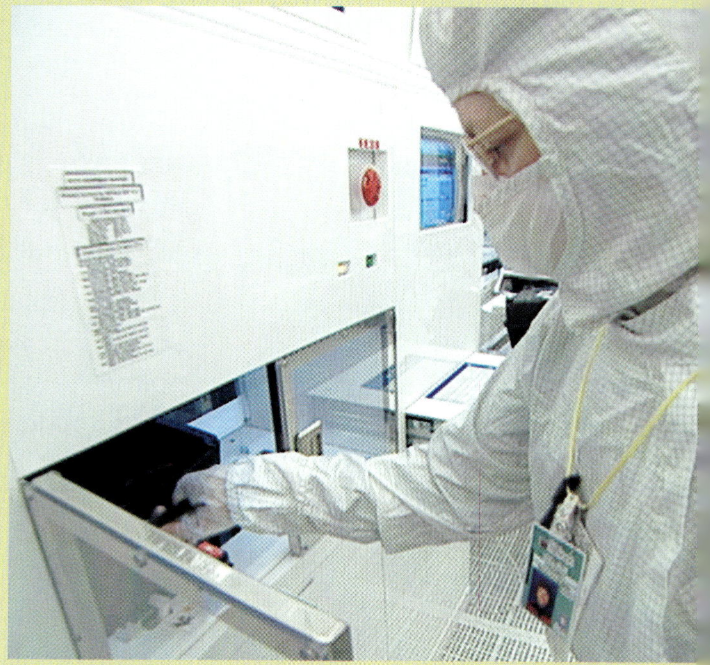

What Happens After *Science in a Technical World*?

You may decide, after completing the *Science in a Technical World* program, that you would like to learn more about the particular industries that are the focus of the different modules. There are a number of programs that prepare people to work in science technology professions. You might want to investigate these, or contact industries directly to find out what training opportunities they offer.

Whatever you decide, *Science in a Technical World* will give you the knowledge and skills to make an informed choice about careers you may never have considered before. It will also give you a wide breadth of skills that you can use in other science-related fields.

Now, welcome to *this* module, which focuses on the semiconductor manufacturing industry.

watch for these symbols

Look for these special symbols as you are reading this book. They have been included to alert you to important learning opportunities, activities, and safety precautions.

? Key Question Consider Key Questions carefully. These questions are designed to help you to build your understanding of the concepts, processes, and skills that are central to the focus industry of this *Science in a Technical World* module.

Tips and Hints Tips and Hints offer ideas that will help you along the way in your investigations. They may assist you in answering questions, setting up laboratory equipment, solving problems, or pulling ideas together.

Notebook Entry This symbol reminds you to stop and make an entry in your laboratory notebook. This could be a record of a laboratory procedure, data, an answer to a question, a conclusion, or other important information.

Boldface Terms Definitions are available for terms set in boldface type in the text. They can be found in the Glossary at the end of the module.

MULTIMEDIA SYMBOLS Each *Science in a Technical World* module includes an introductory video, a multipurpose CD-ROM, and suggestions about industry information that can be found on the Internet. The following symbols alert you to opportunities to supplement your laboratory and classroom activities with one of these multimedia resources.

CD-ROM This symbol tells you to go to the module's CD-ROM and access one or more of its sections. The CD-ROM might include an Introduction, Technician Tasks, Practice, Assessment, and Reference sections.

Video When you see this symbol, you can watch the video to learn more about the production process of the module's focus industry and the role played by technicians in that process.

Internet This symbol indicates that information that will help you in your investigations is available by accessing the Internet.

SAFETY SYMBOLS Safety precautions are extremely important in laboratory work. Be alert to the safety symbols associated with each investigation in this module, and always follow the safety rules they represent.

 **Safety Goggles** This symbol reminds you to wear safety goggles during the laboratory investigation to protect your eyes from chemical splashes or sudden impacts.

 **Laboratory Apron** This symbol reminds you to wear a laboratory apron during the investigation to protect your clothing and skin.

 **Hot Object or Flame** When you see this symbol, be alert to occasions when you might encounter a hot object (such as a hot plate) or an open flame during the investigation. Secure loose sleeves and long hair before starting your laboratory work.

 **Poisonous Substance** This symbol warns that you will be working with substances that should not be tasted. No substance should ever, under any circumstances, be tasted in a classroom laboratory setting.

 **Sharp Objects** This symbol warns you of a cutting hazard. Be extremely careful when using sharp tools. Never handle their cutting edges, and be sure to store them in an appropriate covering or case when not in use. Also, handle laboratory glassware with care to avoid breakage and the danger of getting cut.

 **Shock** This symbol warns you that you are working with materials with an electric shock hazard. Keep the work station dry. Get your teacher's approval before applying a power source to a circuit.

 **Hygiene** This symbol reminds you to wash your hands carefully after a procedure or at the end of the laboratory, as directed.

using your multimedia

Each *Science in a Technical World* module comes with its own video program and CD-ROM.

The video program takes you on a tour of the module's focus industry. You will get to see what products are made, what raw materials are used, and what machinery is important in the industrial process. The video also shows what technicians do in the industry, so you can compare their work with what you do in the laboratory. You may see the complete video at the beginning of the module, or your teacher may show certain parts at appropriate times.

The CD-ROM is a resource for you to use throughout the module. You can access its sections from the main menu by following the voice directions and clicking on the proper screen icons.

The CD-ROM has the following sections:

The **Overview** provides you with information about the focus industry, its products, its processes, and its technicians.

Technician Tasks gives information about what technicians do in the industry. These tasks can include sample preparation, sample checking, sample storage, testing samples, and recording results.

Assessment allows you to test your knowledge and proficiency of the tests, terms, procedures, equipment, and calculations that technicians need to know in the industry.

Practice Exercises gives you an opportunity to simulate some of the tests that technicians do in the industry, expand your knowledge of terms and equipment, and manipulate data sets.

Pull-down menus and icons launch CD-ROM tools:

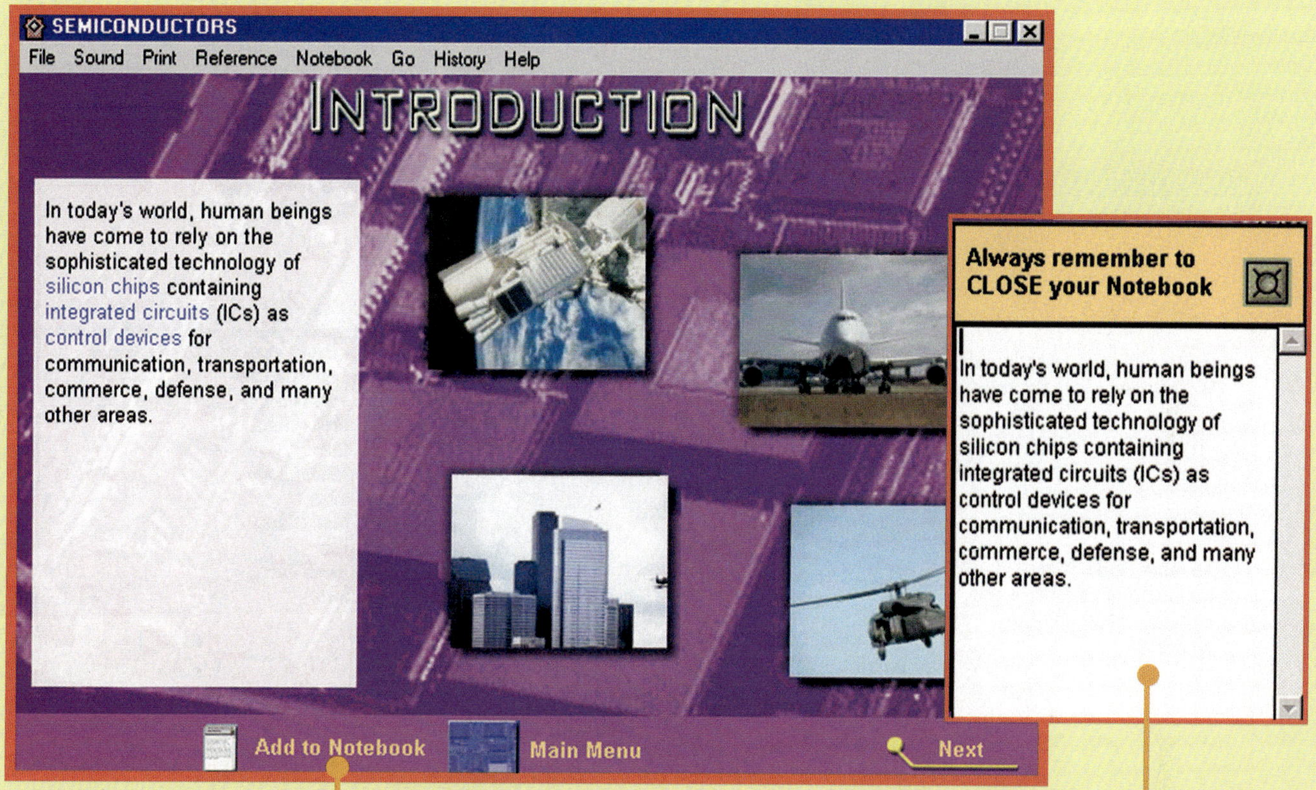

Clicking **Add to Notebook** will copy the text from the current screen to a special CD-ROM notebook.

Your **Notebook** opens from a pull-down menu. You can edit, save, and print the contents of your notebook. Your teacher may ask you to use your notebook when you are doing the Assessment portion of the CD-ROM.

VCR CD-ROM Watch for **Multimedia Links** throughout this book. They suggest what part of the video or CD-ROM will be most useful to you at each stage of your work with this module.

History tracks where you've been on the CD-ROM. You can navigate to a previous screen by clicking on its name in a displayed list of screens.

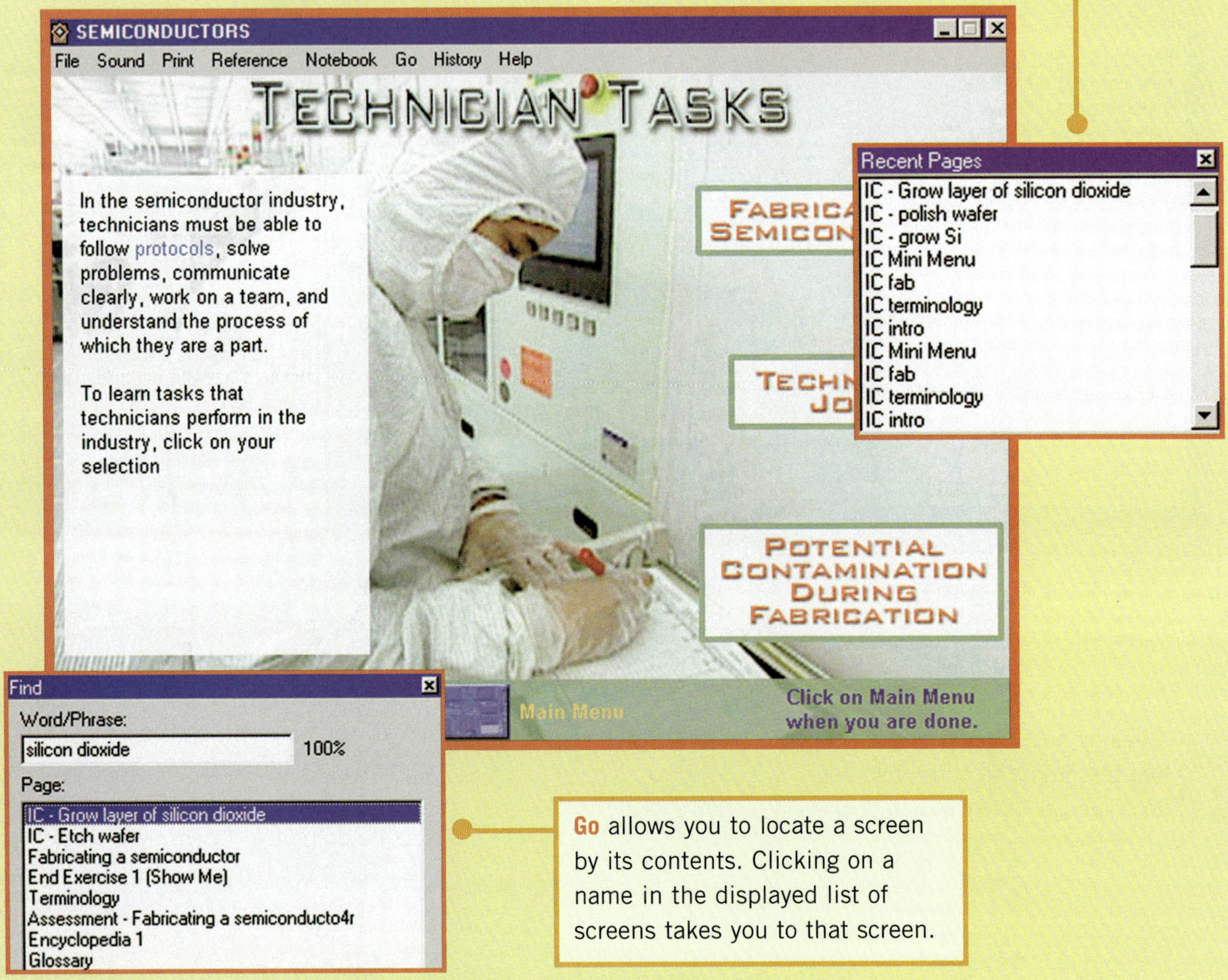

Go allows you to locate a screen by its contents. Clicking on a name in the displayed list of screens takes you to that screen.

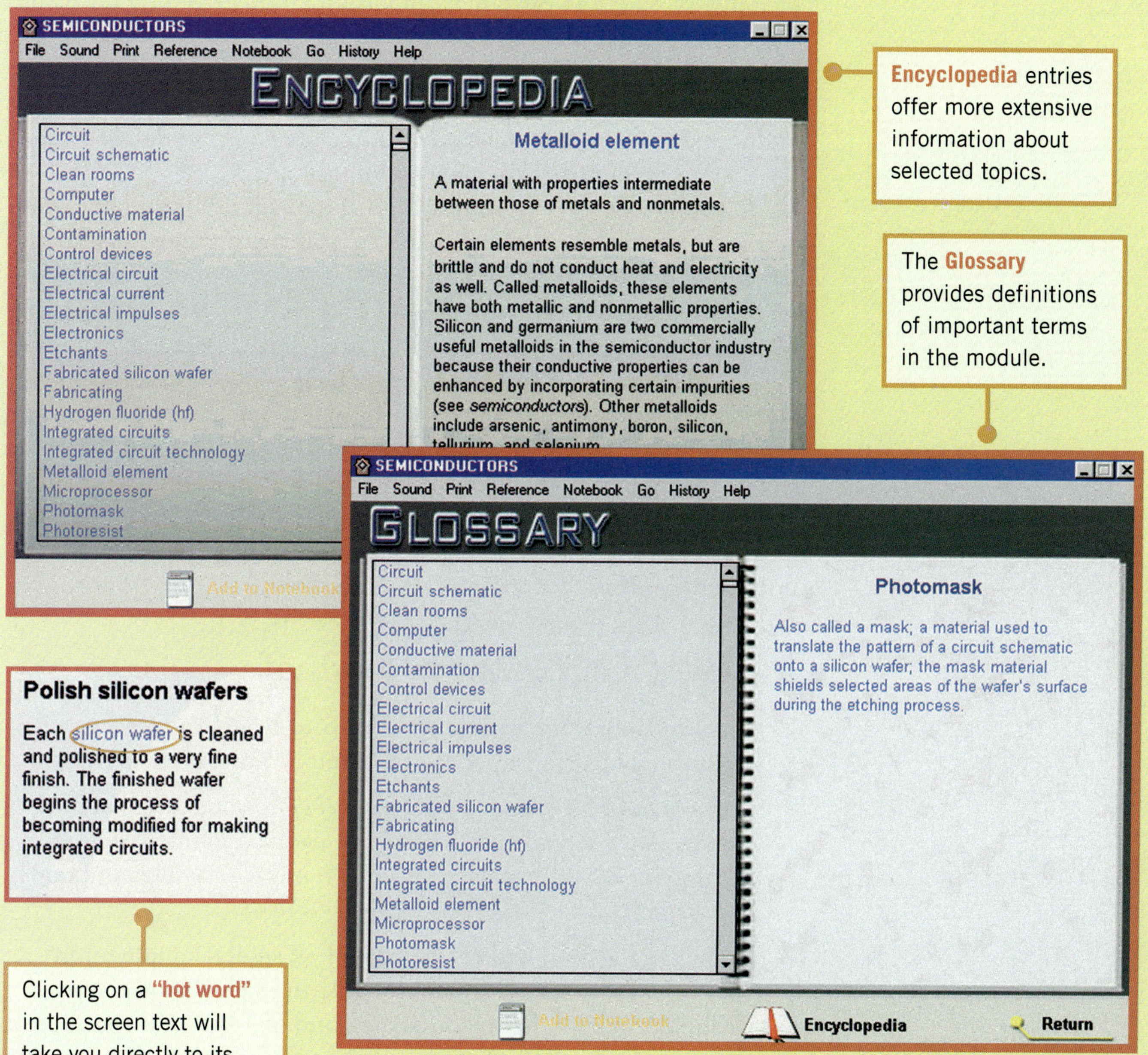

Encyclopedia entries offer more extensive information about selected topics.

The **Glossary** provides definitions of important terms in the module.

Polish silicon wafers

Each silicon wafer is cleaned and polished to a very fine finish. The finished wafer begins the process of becoming modified for making integrated circuits.

Clicking on a **"hot word"** in the screen text will take you directly to its Glossary definition.

Watch for **Multimedia Links** throughout this book. They suggest what part of the video or CD-ROM will be most useful to you at each stage of your work with this module.

safety in the laboratory

In industry, safety is extremely important. Industries rely on health and safety guidelines set by OSHA, the Occupational Safety and Health Administration, U.S. Department of Labor. These guidelines are tailored to particular industrial processes and practices. In some industries, for example, visitors to a facility must undergo a safety orientation before they are allowed into the plant.

You will have the opportunity and responsibility throughout your *Science in a Technical World* work to observe standard health and safety practices in the laboratory. These practices and guidelines are for your own good; they will ensure that your investigations are rewarding and safe experiences for you and your classmates.

Each laboratory in this book has a Safety section. It is very important that you read and understand the safety precautions before you begin any lab. If you don't understand the safety rules, ask your teacher to explain them. The Rules of Laboratory Conduct listed below apply to *all* laboratory activities included in the *Science in a Technical World* program. Read them now and review them periodically until they become second nature.

Rules of Laboratory Conduct

- Perform laboratory work only when your teacher is present. Unauthorized or unsupervised laboratory work is not allowed.
- Your concern for safety should begin *before* you start a laboratory activity. Always read and think about each assignment before starting it. Ask your teacher to explain any safety rules you don't understand.
- Know the location and correct use of all safety equipment in your laboratory. These should include the safety shower, eye wash, first-aid kit, fire extinguisher, and blankets. Know the location of all exits and proper evacuation routes.
- Wear impact/splashproof safety goggles for all laboratory work.
- Make every effort to protect bare skin. Shorts or short skirts must not be worn; long pants and shirts with sleeves provide better protection. Wear closed shoes (rather than sandals or open-toed shoes).
- Restrain loose articles of clothing, jewelry, and hair.
- Clear your laboratory bench of all unnecessary materials, such as books or articles of clothing, before starting your work.

- Be extremely careful when using sharp tools; never handle their cutting edges, and be sure to store them in an appropriate covering or case when not in use. Also, handle laboratory glassware with care to avoid breakage and the danger of getting cut.
- Check chemical labels twice to make sure you have the correct substance and the correct concentration of a solution (some chemical formulas and names can differ by just a single letter or number). Read and follow the appropriate safe-usage guidelines for each chemical.
- You may, at times, be asked to transfer laboratory chemicals from a common bottle or jar to your own test tube or beaker. Do not return excess material to its original container unless authorized by your teacher.
- Avoid unnecessary movement and talk in the laboratory.
- Never taste laboratory materials. Do not bring food, drinks, or chewing gum into the laboratory. Do not put fingers, pens, or pencils into your mouth while in the laboratory.
- Wash your hands before leaving the laboratory!
- If you are instructed to smell something, do so by fanning some of the vapor toward your nose. *Do not* place your nose near the opening of the container. Your teacher will show you the correct technique.
- Never look directly down into a test tube or beaker; view the contents from the side. Never point the open end of a container toward yourself or a neighbor. Never heat a container directly in a Bunsen burner flame.
- Any laboratory accident, however small, should be reported immediately to your teacher.
- If a chemical spills on your skin or clothing, quickly rinse the area with plenty of water. If the eyes are affected, washing with water must commence immediately and continue for at least 15 minutes. Professional assistance must be obtained.
- When discarding or disposing of used materials, carefully follow the instructions provided.
- Return equipment, chemicals, aprons, and protective goggles to the designated locations.
- If in doubt about a procedure or precaution, *ask!*

Module Focus

In this module, you will be working in teams to solve a problem faced by technicians who fabricate semiconductors.

Problem to Be Solved

What new training procedures can increase the background knowledge, team ethic, and communications skills of a semiconductor technician?

introduction

The Problem

Stratus Semiconductors, a company on the East Coast of the United States, has recently merged with Vertical Systems, a company based in the Far East. Vertical Systems brings a somewhat different workplace system to the merger; the company trains its technicians in a variety of tasks and ensures that the technicians understand the science and technology behind the semiconductor manufacturing process. Vertical develops its strong sense of teamwork by educating each team member, including scientists, engineers, and technicians, about all of the roles that contribute to the semiconductor fabrication process.

Stratus has agreed to alter the design of its technician training program to meet the teamwork expectations of Vertical Systems. Although Stratus has always used a team approach in its manufacturing process, the technicians on each team were not fully informed about all of the science and technology behind their tasks. They were skilled in performing the standard procedures unique to each role. Stratus technicians were trained to do one task at a time and received training on a new task only when they assumed new responsibilities.

Over the years, however, Vertical learned that technicians who were trained in more than one task and who understood clearly how each task contributes to making a product that meets specifications have a greater sense of ownership of the product. A greater sense of ownership, Vertical learned, leads to greater employee job satisfaction as well as an increased product yield for the company.

In this module, you will be working as one of a group of semiconductor technicians at Stratus. You will be one of the first in the Vertical Systems training program. As you engage in the training, it will be your job to provide feedback to the training designers. The final challenge will be to build and test a small robot to demonstrate your team's knowledge of the semiconductor industry, its skills in performing workplace tasks, and its ability to work together as an effective team.

Multimedia Links

CD-ROM

Some of the terms introduced in the description of the problem facing you as a semiconductor technician may be unfamiliar. You can look them up in the CD-ROM Glossary and Encyclopedia, as well as the glossary at the back of this book.

Your Role in Solving the Problem

Your role as a technician in the newly merged semiconductor company is to follow the procedures to learn as much as you can about the semiconductor manufacturing process and to build a team ethic with your fellow technicians.

Your training will involve some of the following science and technology concepts:

- how a wafer, a chip, and an integrated circuit are defined and related to one another
- how a circuit works
- how solid-state electronic components such as diodes and transistors work
- how an integrated circuit works
- how a chip is made
- how an assembly line process works, and the importance of each step to the final product
- how robots can be programmed to serve a purpose in an industry

Although semiconductor protocols (set procedures) are usually conducted under the cleanest conditions possible, it would be impossible to make your classroom as clean as a semiconductor-industry "clean room." Therefore, in this training you will gain knowledge of fabrication protocols through simulations and modeling.

You will be using the technical workplace model of *teamwork, specialization,* and *collaboration* as you engage in the semiconductor training. Throughout the course of this module, you will be working in teams of three or four students. You will be depending on every member of your team to complete the work on time and to verify all results.

At the end of your training, you will put your knowledge to work by cooperating with your team in making a small robot. This product will incorporate the electrical components you learn about in your training. After testing your robot, you will suggest industrial applications for similar devices in the workplace. Finally, you will work with your team to review and revise the training procedures for manufacturing chips for your company.

industry overview

Fact Sheet on Semiconductors

In today's world, people have come to rely upon the sophisticated technology of silicon chips (integrated circuits, or ICs) as control devices for communication, transportation, commerce, defense, and many other areas. These chips, small enough to fit on the tip of your finger, are able to store, transmit, and interpret information so rapidly that they have changed the way people communicate with each other and interact with their world. But computers were not always so fast or even close to being so small. The first computer, unveiled at the University of Pennsylvania in 1946, filled an entire room and required 18,000 vacuum tubes to direct the current pathways. Like light bulbs, the vacuum tubes burned out so often that computer crashes were an almost daily occurrence. And, even on a good day, the computing power of this early model scarcely approached the power of a modern hand-held calculator.

Today's computer runs on thousands of complex integrated circuits applied layer by layer to thin wafers of silicon. The science and technology that goes into the design of the silicon chip combines physics, chemistry, engineering, and electronics. The element silicon is an example of a semiconductor—a material that is intermediate between being a good conductor of electricity (like copper) or a non-conducting insulator (like glass) at room temperature. But silicon is a semiconductor whose conducting properties can be manipulated by adding other elements (integrating them) to its diamond-like crystal structure. Applying the right materials in the right patterns to make miniature integrated circuits involves the skilled work of the semiconductor technician.

This module will allow you to explore what semiconductors are, how they are used to make integrated circuits, and how these integrated circuits are essential to a variety of products. You will learn the role that technicians have in all of these processes.

Background: Making an Integrated Circuit Based on Silicon, a Semiconductor

A circuit is a pathway along which millions of moving electrons stream to form an electrical current. Electrons flow easily through a conductive material such as a

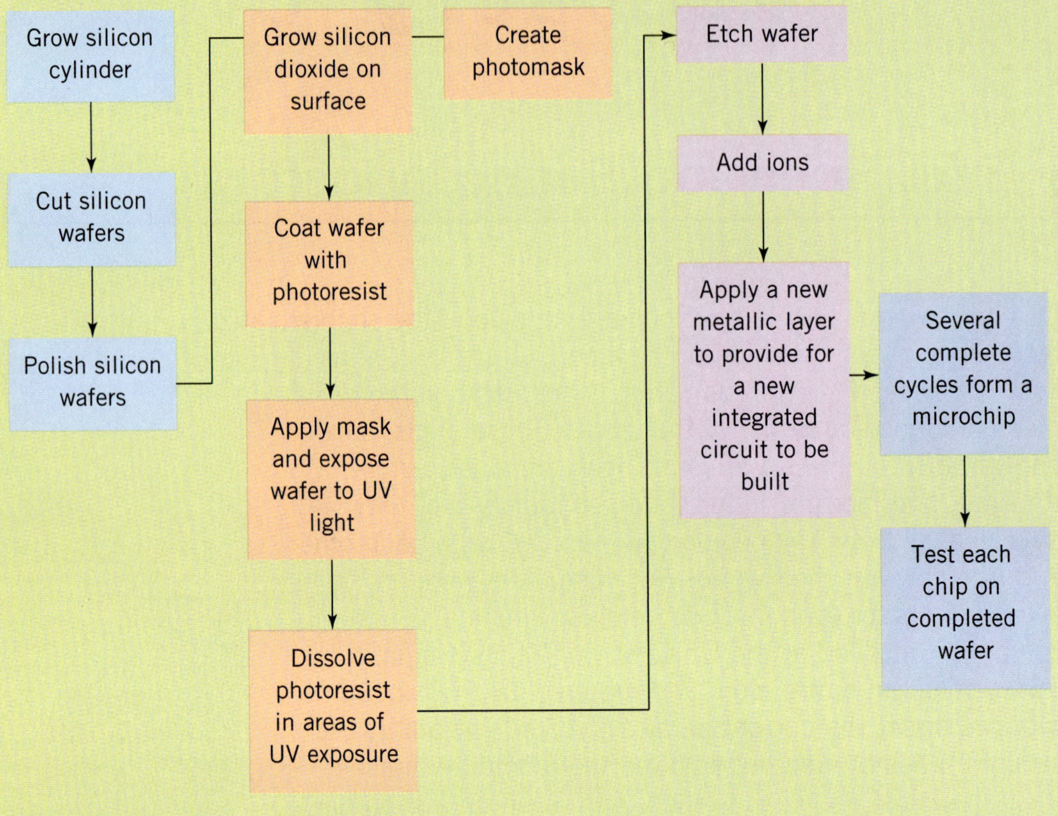

Figure 1
Fabricating a semiconductor.

metal (copper, silver, gold, aluminum) and less easily through a semiconductor. But a semiconductor is a material whose conductive properties can be altered. During the fabrication of a semiconductor (Figure 1), additional atoms are embedded in silicon; the atoms selectively alter the conductivity of certain regions. The pattern by which these atoms are embedded affects the pathway of the electron flow.

A semiconductor is the basis of every integrated circuit. Here is the step-by-step process for making one:

Step 1:

The process of making (fabricating) an integrated circuit takes place in a facility known by the nickname of the "fab." The fab has its own water treatment facility (Figure 2), as well as the capacity to super-clean its air (Figure 3). Even so much as a single speck of dust can ruin an integrated circuit.

Step 2:

Pure silicon is heated to a temperature of 1500 °C. A tiny seed crystal of silicon is rotated through the molten material, initiating the growth of a large, pure crystal in the shape of a cylinder. The cylinder is then cut into a series of thin wafers. By the end of the process, a single wafer will produce hundreds of identical chips.

Step 3:

Each silicon wafer is cleaned and polished to a very smooth finish. The finished wafer then enters the process of building an integrated circuit.

Figure 2
Part of the water-treatment facility at the fab.

Step 4:

A circuit schematic is designed that will result in an integrated circuit useful for a specific purpose. That schematic is then translated into a series of stencil-like materials called photomasks. They will be used to translate the pattern of the schematic onto the silicon wafer in a series of layers.

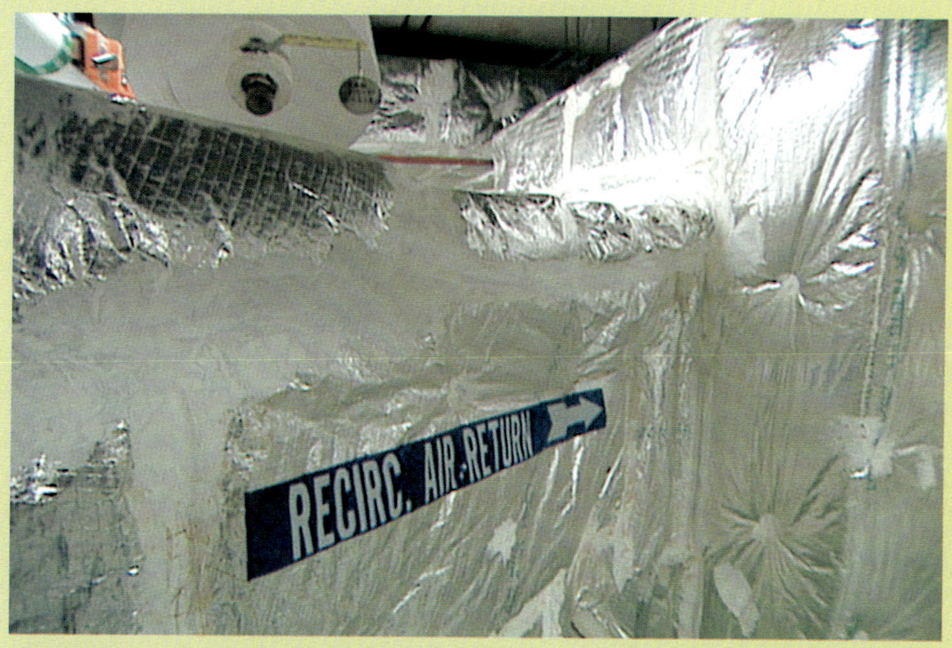

Figure 3
Part of the air-treatment facility at the fab.

Step 5:

A layer of silicon dioxide is grown on the surface of the wafer by exposing it to extreme heat and gas.

Step 6:

The wafer is treated with a substance called photoresist. The photoresist ensures that the pattern of the photomask will transfer exactly to the wafer. A mask is applied, and the first photoimage is developed with UV light. As the unmasked areas are exposed to UV light, they become sticky and are easily dissolved.

Step 7:

The wafer is treated with a solvent, which dissolves the sticky, UV-altered photoresist. The gases remove the layer of silicon dioxide that was grown on the wafer's surface (step 5) etching a pattern into it. Only the developed, unmasked areas of the wafer present bare silicon for the next step. Undeveloped areas remain coated with undeveloped photoresist.

Step 8:

In a process called "doping," metal atoms are embedded into the etched pathway—called a crystalline lattice—of the bare areas of the silicon wafer to make zones of selectively conductible material. (The areas stilled covered with photoresist are not affected.) In subsequent steps, layers of metals may be laid down to form electrical connections between the doped areas.

Step 9:

The remaining photoresist is removed from the undeveloped regions of the surface, and the wafer is ready for the next round of treatment.

Step 10:

The process of coating with photoresist, photomasking, etching, doping, and applying metals is repeated. The result is an average of 20 such layers in the hundreds of rectangular chips on the wafer.

Step 11:

When the wafer is complete, the chips are separated to be individually examined and tested. Because the crystal lattices are so minute, even the smallest speck of dust in the layering mix can ruin a chip. For this reason, the entire assembly process for these miniature integrated circuits takes place in **clean rooms**—rooms in which the air is so clean that not even the tiniest particle of dust or organisms such as bacteria and fungi are present. People who work on the wafers, including technicians, perform their tasks wearing absolutely clean, protective clothing, called "bunny suits."

Developing Your Skills Base

The semiconductor module is designed to help you develop *laboratory and problem-solving skills* that will be useful to you throughout the *Science in a Technical World* program, and throughout your lives. You will have many opportunities to practice and refine these skills in the laboratory and through exercises on the semiconductor CD-ROM. It will be your responsibility to monitor how well you are mastering a particular skill, and to get help when you think you need it.

Pay special attention throughout the module to how you observe, measure, and test for the properties of different materials. The ability to make ***quantitative and qualitative observations*** *is a vital skill for laboratory technicians.*

Problem solving is the most essential skill for any technician. Learning to use problem-solving strategies (science processes) will help you solve the semiconductor problems (and other science problems in the future) as efficiently and effectively as possible.

important concepts, processes, and skills

Science and Technology Concepts

When you finish this module, you should be able to define and give examples of

- electrical circuits
- integrated circuits
- conductivity
- resistance
- transistors
- semiconductors
- diodes
- doping
- wafers
- chips
- robots
- photolithography
- etching
- super-clean environments

Science Processes

When you finish this module, you should be able to

- *define* problems to be investigated
- *use* models to investigate science and technology problems
- *analyze* models used to investigate science and technology problems
- *gather* data
- *record* data
- *analyze* data
- *arrive* at conclusions based on data analysis
- *communicate* results

Laboratory Skills

When you finish this module, you should be able to

- *follow* protocols for experimental procedures
- *wire* a simple circuit
- *make* electrical measurements using a multimeter
- *make* a circuit on a breadboard using solid-state components
- *insert* and *operate* transistors in a complex circuit
- *make* two models of steps in the photolithography process
- *use* a microscope
- *assemble* and *test* the performance of a robot
- *use* appropriate personal safety equipment

Workplace Skills

When you finish this module, you should be able to

- *conduct* research, using the Internet and other sources
- *read* and *understand* laboratory procedures
- *access* and *use* online resources
- *work* cooperatively with team members
- *recognize* the importance of your work as a member of a manufacturing team
- *evaluate* a training program
- *follow* correct safety procedures

Defining the Problem

The problem that you face as a technician in the new training program is to provide clear and useful feedback on the usefulness of the program in achieving the following goals:

- understand the science and technology of the semiconductor manufacturing process
- develop a sense of ownership of the product
- become an integral member of a team that works together

After each training step, the leader will provide you with a form on which you can provide feedback on how well the step contributed to the goals of the training. You will need to work with your team members to communicate feedback that will be most useful to the designers of the training session.

The Importance of the Training

Your training in semiconductor science and technology will be similar to that of actual technicians in this line of work. Some manufacturing facilities send their technicians to local two-year colleges for training, others send technicians to similar manufacturing facilities, and still others receive their training on-site in their own plants. As your training on semiconductors begins, you will encounter familiar as well as unfamiliar concepts. In an actual workplace, there are usually people to assist you in gaining skills and understanding; in a *Science in a Technical World* module, there are many resources available to find the answers to questions you may have.

Some of the resources available to you are

- the *Making Semiconductors* video, about the work of semiconductor technicians and the importance of that work in making a product that meets specifications
- printed materials with valuable background information and planning tools.
- a variety of important Web sites and databases that offer both data and background information for solving problems
- the *Making Semiconductors* CD-ROM, useful for overview, practice, and glossary and encyclopedia assistance. You will find that you can use the CD-ROM at your own pace and as frequently as necessary to build your skills and understanding.
- your fellow technician trainees

Throughout your training, you will keep important records. In your laboratory notebook, you will record procedures followed and data collected for all laboratories. In the following activity, you will set up a laboratory notebook for recording your work throughout the technician training.

Becoming Familiar with Semiconductors

Activity Overview

Your technician training will begin with a brief orientation session designed to provide the background necessary to understand some of the science and technology behind semiconductor manufacturing.

Step 1 Setting Up and Maintaining a Laboratory Notebook

Materials

- bound notebook, preferably with graph paper pages
- pen

Legal Issues

Laboratory notebooks are used to support applications for patents and to document all laboratory tests done for a company. To be legally acceptable, laboratory notebooks must be completed and signed by technicians daily. Each page must be

- numbered
- dated
- witnessed by another person

It is very important that your laboratory notebook is neat, accurate, and written in pen, because others may need to repeat your experiment or procedure. Pages should be dated so that it is easy to follow the sequence of the experiments or tests. Also, any unique or new event should be recorded, as well as the exact time the event took place.

Procedure (for 3 to 4 team members)

Each team member should keep his/her own notebook, containing a record of

- tests and procedures
- predictions
- data
- observations
- calculations
- notes
- conclusions

You may also be required to include a description of each complete test and a summary of tasks completed by your team.

At the end of each class period, remember to have your notebook signed by another member of your team.

Handling Data

Data must *always* be labeled and described. Be sure to record

- the time of day the test was run
- appropriate environmental conditions at the time the test was run
- your observations about the samples

A table is always a good way to present data, as it is easy to review later.

Connecting to the Problem

Although you will be very familiar with a procedure and its results are very familiar while you are running a test, later it can be hard to remember just what a sample looked like or what results the test produced. This is why a well-kept laboratory notebook is necessary. A finished product that meets specifications can depend upon your test results, so the laboratory notebook provides a record of how the tests were run in the laboratory, serving as a history for each test and procedure. No information is unimportant.

Only clear, accurate data and directions will do for your laboratory notebook.

Multimedia Link

CD-ROM

In addition to your laboratory notebook, there is an electronic notebook on the CD-ROM. You can copy text from the CD into this notebook, edit the text, print it, and save it to a floppy disk.

Step 2 Researching Semiconductors

Activity Overview

To begin learning about the basics of semiconductors, you and your teammates will generate questions about semiconductors, then conduct research to find answers to those questions. Be sure that you keep your question list throughout the training session, as you will have many opportunities to find answers to your questions in a variety of ways.

Materials

- semiconductor CD-ROM
- *Making Semiconductors* video
- Web sites
- print materials on semiconductors

Safety

There are no special safety requirements needed for this activity.

Procedure **(for 3 to 4 team members)**

1. Read over the Industry Overview on Semiconductors (pp. 3–6); then work with your team to make a list of any questions that you have. Discuss your questions, and prioritize your list from most to least important for understanding how semiconductors work.

2. Here are some questions that you may want to consider in your research:

- What are semiconductors?
- Why is silicon used in making semiconductors?
- Why does the semiconductor fabrication process have to take place in such a clean environment?
- What is the function of water in making semiconductors?
- For what are semiconductors used?
- What is the difference between a wafer and a chip?
- What is the role of robots in the semiconductor fabrication process?

3. Watch the video to see how many of your questions it answers. Divide the responsibility for certain questions among your team members, and set a deadline for the research effort.

4. The CD-ROM glossary will provide vocabulary definitions, which, in turn, may link to more in-depth explanations in the encyclopedia. You may wish to print out the glossary searches you make in this first session and add them to your notebook.

5. Web sites provide valuable resources. Make a list of key words for your search, such as semiconductors, silicon chips, and integrated circuits. Record the Web sites you find so you can share the information with the other trainee technicians in your group.

Analyzing the Data

Collect the answers to all of your questions and exchange your findings with other groups in your class. To share information, your class may decide that to set up a Web page or publish a printed or online newsletter is an efficient way to communicate.

Review all of the information you found on semiconductors.

- Were there any areas of disagreement in the information you found? If so, what were they and how can they be resolved?
- What were the most useful and reliable sources of information you used? How could you use these sources as you progress through the training?
- What questions are still unanswered? Where might you might find this information?

Arriving at Conclusions

After collecting your information, discuss the amount of information available on semiconductors. Did you find much more than you needed, or was the information difficult to find? Was most of the information too technical for your understanding? Not technical enough? Make a file all of the information so you can access it easily during the module. ? note

Connecting to the Problem

Evaluate how this first part of your training session helped you understand the science and technology behind semiconductors and their fabrication.

? *How well did this research exercise help you understand semiconductors?*

? *What activity would have been more useful in helping you learn more?* note

Provide feedback to the training staff on the survey form from your teacher. You will be asked to evaluate your teamwork during this step of the orientation.

Step 3 Examining a Chip

Laboratory Introduction

In this first investigation, you and your teammates will take a close look at a silicon chip.

Background

To begin your experiences with integrated circuits, you will examine a chip at close range under a dissecting microscope. Then you will make a drawing and describe as much of the chip as you can. You will later relate this image to that of a circuit you build. You will also compare the chip to the greatly magnified image on the CD-ROM.

Materials

- stereomicroscope with at least 30x magnification
- chip sample
- drawing paper and pencil
- CD-ROM

Safety

No special safety precautions are needed for this investigation.

Procedure (for 3 or 4 team members)

Examine the chip, first with the naked eye, then under the lowest, then the highest magnification of the dissecting microscope. Draw what you observe at each step. Make a table like Table 1 in your notebook to record your observations.

A silicon chip without its casing.

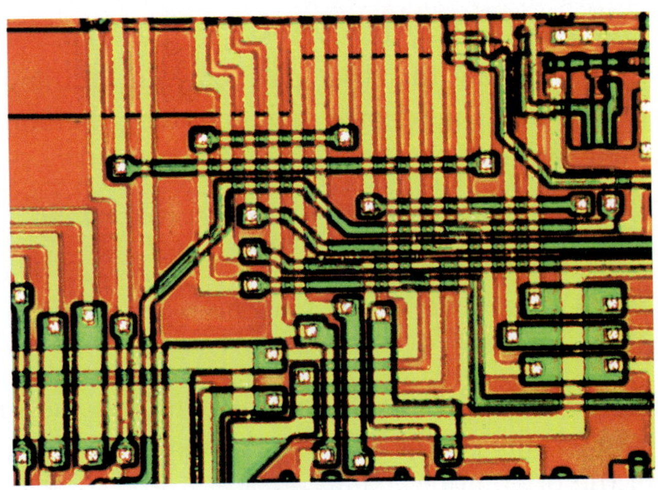

A silicon chip, magnified.

Table 1

Observations of a Silicon Chip	
View	**Observations (words and drawing)**
Naked Eye	
Low Magnification (10x)	
High Magnification (30x or more)	

Study the parts of the chip you included in your drawings. What do you think are the functions of the parts? Record your answers in your notebook, and check your ideas with information on the CD-ROM.

Multimedia Links

CD-ROM VCR

Review the video *Making Semiconductors* to see the parts of a chip. As well as you can, identify the parts of the chip from the information presented on the video, and label your drawing in the table. Look up any unfamiliar terms on the CD-ROM Glossary and Encyclopedia.

Analyzing the Data

- After examining the chip under different magnifications, record your impressions of the complexity of the chip.
- From what materials did it appear that your chip was made? How do you think these materials contribute to the working of the chip?

Arriving at Conclusions

Chips provide the means for processing large amounts of information.

What visible features suggest the ways that chips do this?

Connecting to the Problem

As a trainee technician in semiconductors, you are asked to provide feedback to the training designers about the effectiveness of each step in building your understanding of semiconductors.

What did you learn from this activity about the construction of a chip?

How did the CD-ROM help you understand what the parts of the chip were?

What could have been added to the activity to help you understand more about the chip's construction? Complete the survey form provided by your teacher. note

Laboratory 1 Wiring a Simple Electric Circuit

Laboratory Introduction

In this investigation, you will learn how an electrical circuit works and how to build a circuit. This understanding is the basis for a clear understanding of how more complex integrated circuits work.

Background

When you think of the word "circuit," think of "circle." A circuit is a closed system (circle) in which there is an energy source (electricity, in this case) and a conduit or path for the electricity that leads back to the source. You will begin this investigation by designing, drawing, and making a very simple complete circuit and will learn some of the symbols that electrical engineers use to represent its various parts. You will then experiment with making breaks in the circuit. In the following training step, you will use your knowledge to make a more sophisticated circuit—one capable of integrating more components than a simple circuit.

Materials

- alligator clips
- coated wires, as needed, 15-cm lengths
- wire strippers
- battery holders and clips
- 1.5V dry cells with spring-type terminals
- flashlight bulb in holder
- switch
- drawing paper
- buzzer or small plastic propeller

Safety

- Do not touch bare wires. They may get hot.
- Only use a 1.5V dry cell for the simple circuit.

- Use caution when handling the wire strippers. Always strip the insulation away from yourself.

Procedure (for 3 to 4 team members)

1. Your teacher will demonstrate how to use a wire cutter/stripper to cut a 15-cm piece of wire and how to strip about 0.5 cm of insulation from each end. You will prepare these pieces as needed for your circuit.

2. Examine the dry cell, holder, clips, wires, bulb, and bulb holder. On paper, work with your team members to design a circuit in which electricity stored in the dry cell can pass through the wires, through the bulb, through the wires again, and back to the dry cell. Check your drawing of the circuit with another group's. Help each other to spot flaws in the drawings.

3. When you are satisfied that your circuit will work successfully, use the materials provided to build the circuit and test it out.

? *How will you know that your circuit is successful? Record your answer in your notebook.*

4. Exchange diagrams with another group, and ask that group to build a circuit from your diagram. Remember that a good diagram is one that others can interpret.

5. Your teacher will provide you with symbols used by electrical engineers in circuit diagrams. Redraw your diagram using these symbols. Again ask another group to check your work for accuracy and clarity.

6. Find out what happens when you disconnect one of the connections to make a break in your circuit. Observe what happens to the light, and record your observations in your notebook. Insert a switch into the circuit, and demonstrate how to use the switch to break and restore the circuit.

7. Working with your group, figure out how you could use your complete and incomplete circuits to send a message to a group across the room. Share your plan with a nearby group. Decide what information the team across the room would need to interpret your message successfully.

8. When you think your plan will work, put it into action and send your message. Find out if the other group interpreted it correctly.

9. Obtain a buzzer or a small plastic propeller, and wire it into your circuit in place of the light bulb and holder. When you close the switch, what happens to the component?

Analyzing the Data

Review your complete and broken (incomplete) circuits.

- What are the necessary components of a complete circuit?
- How is breaking the circuit a useful action? How does the switch help manage this?
- Were you able to communicate a message using a complete (light on) and broken (light off) circuit? What problems did you have?
- What were the advantages in making the circuit diagram? How did the shorthand symbols help in this task?
- What happened when you included a buzzer or a propeller in your circuit?

Record the answers to the questions in your laboratory notebook.

Arriving at Conclusions

Based on your investigation, describe the basic components of a complete circuit. Trace the path of electricity through the circuit. Record your descriptions in your notebook.

Relate your "large scale" circuit to the magnified integrated circuit you worked with in laboratory one.

What similarities are there? What differences?

Connecting to the Problem

Your fabrication facility is particularly concerned that everyone in your team of technician trainees understands

- circuitry
- making and interpreting diagrams
- working as a team

Discuss with your team how this investigation helped you advance in each of these areas. Record the main points of your discussion in your notebook. Complete the survey form provided by your teacher.

Laboratory 2 Controlling the Flow of Electric Current

Laboratory Introduction

In this investigation, you will learn some basic characteristics of electricity and how electronic components are used to control its flow.

Background

In Laboratory 1 you made an electrical circuit. What was moving through the circuit? The answer is electric current, which is the flow of electrons. Electrons—basic subatomic particles that surround atomic nuclei—are fundamental to all matter. Even the smallest object that you can see contains trillions of electrons.

Electrons can be at rest or they can move. When they move, we call it **electric current.** Some materials allow electrons to flow while others do not, a characteristic that will be discussed in the next laboratory. Substances that allow electrons to flow are called **conductors,** while those that don't are called **insulators.** There is actually no such thing as a perfect conductor. Even in the best conductors, the conductor's particles tend to get in the way of the flowing electrons, resulting in a measurable property called electrical **resistance.** Conductors have low resistance to the flow of electricity, and insulators have high resistance.

Resistance is not always a bad thing in designing circuits. Sometimes we want electricity to flow, and other times we don't. For example, we DO want electricity to flow through our radios, but we DON'T want electricity to flow through us. That is why in some parts of the radio we use conductors such as wires, and in other parts, especially those we are likely to touch, we use insulators like plastic or rubber.

The whole purpose behind the design of electrical devices is to control the flow of electrons—both their direction and amount. There are dozens of different kinds of **electrical components** that can help us do just that. There are components that stop, switch, block, amplify, store, and otherwise control the flow of electrons. In this laboratory we will learn about a few of them.

Resistance is measured in units called ohms, which can be measured with a multimeter, a tool that can make many different kinds of electrical measurements. Your teacher will demonstrate the basic operation of your multimeter (Figure 4). A substance that is a conductor allows electrons to flow freely—in other words, the substance has low resistance. The greater the number of ohms measured by the meter, the harder it is for the current to flow through the material.

Materials

- breadboard
- several pieces of insulated wire (6 cm long—using the wire strippers, strip away enough insulation to expose about 0.5 cm of bare wire at each end)
- wire strippers
- red LED (light-emitting diode)

Figure 4
A multimeter.

- 1000-ohm (1 K Ω) resistor
- 9-volt battery, with leads attached
- terminal clip with leads for a battery
- multimeter or volt-ohm meter
- diodes
- CdS photoresistor (also called a photocell)
- ruler
- #2 pencil
- graph paper (or notebook page)

Safety

- Use caution when handling the wire strippers. Always strip the insulation away from yourself.
- Follow directions to avoid shocks.
- Put all materials away when you have finished.

Part A Using a Multimeter to Measure Resistance

Step 1 Measuring Resistance in a Graphite Wire

1. To measure resistance accurately you must adjust the controls on your multimeter. When you touch the probes together, the meter should read zero. If it does not, adjust the meter until it does.

2. Using a straight edge as a guide, draw a 10-cm line on a piece of paper with a #2 pencil. Darken the line by drawing over it with the same pencil 20 times (Figure 5). Then make marks 5 mm apart along the line. This will be the first "wire" that you test.

Figure 5
Drawing the 10-cm line.

3. Touch one of the probes of the multimeter to the point nearest the end of the line. Place the other probe as close as you can to the first one, without actually touching it. What is the resistance reading on the multimeter?

4. Slowly increase the distance between the probes (Figure 6). Notice what happens to the resistance value on the multimeter.

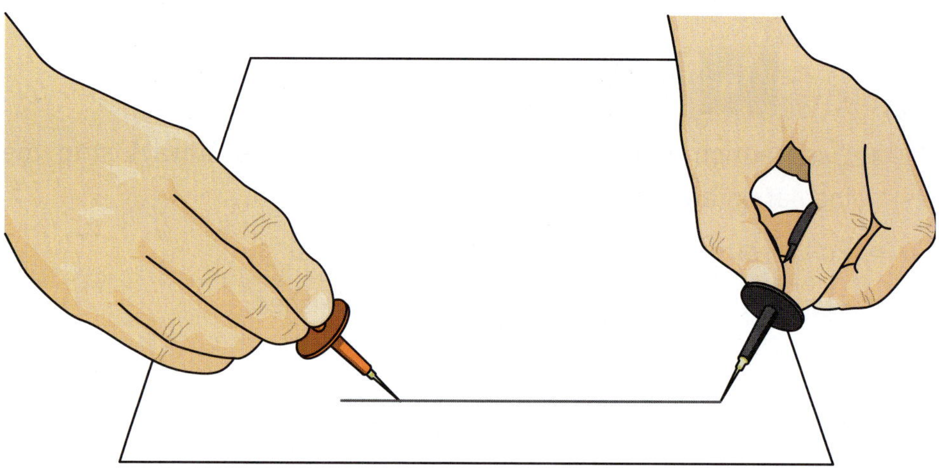

Figure 6
Moving multimeter probes farther apart.

5. Position one multimeter probe on the first point. Measure the resistance as you place the other probe at each of the 5-mm marks, and record the information in your notebook. Then make a graph of

the resistance vs. distance along your pencil wire in your laboratory notebook.

Analyzing the Data

- Describe the shape of the graph of resistance vs. distance.
- In words, describe the relationship between the distance and the resistance.

Arriving at Conclusions

? *Pencil lead is actually carbon in the form of graphite. Is graphite a conductor of electricity?*

? *Does graphite offer any resistance to the flow of electricity?*

? *Why do you think the resistance changed as you moved the probe?*

Step 2 Changing the Resistance of a Graphite Wire

You found that graphite is a conductor but not a perfect conductor. Predict what will happen if you apply more graphite to your pencil wire. Will the resistance readings be less than they were before? The same? Greater? Why? Record your answer in your notebook.

1. Darken your graphite line by applying 20 more strokes with your #2 pencil.
2. Repeat steps 3 and 4 in Step 1.
3. Compare this graph with the one from Step 1.
4. Were your predictions correct?
5. Discuss your findings with your team and record your explanation in your notebook. You may want to try the experiment again with different distances or different amounts of graphite.

By varying the amount of resistance in the circuit, a technician can control the amount of current flowing at any given time. Sometimes resistance is manipulated by varying the length of the pathway as you did in this procedure. Sometimes, as you will see in Part B, the technician simply adds to the circuit a component that will offer a certain amount of resistance to the flow of current.

Part B Using Breadboards and Solid State Electronic Components

Background

A convenient device for building temporary circuits is a *breadboard*. Breadboards are plastic plates perforated with rows of holes for

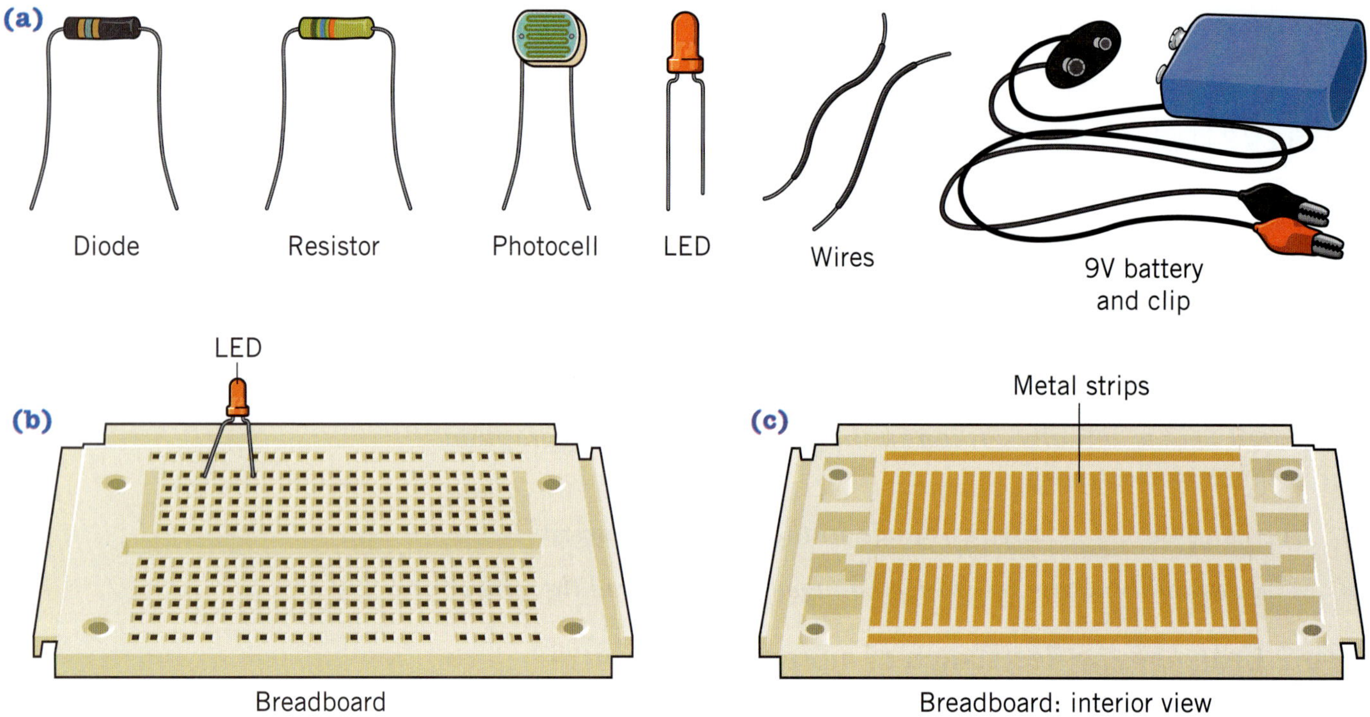

Figure 7
Parts of a breadboard circuit.
(a) Components used in a simple circuit
(b) Components inserted in the breadboard
(c) Metal strips beneath holes in the breadboard

holding and connecting wires and electrical components (Figure 7a). Parts of the circuit are connected on a breadboard by poking wire ends into the holes on the board (Figure 7b). But when planning connections, it is important to know what lies beneath the surface of the breadboard. Just below each short row of holes is a strip of conducting metal. All wires poked into the holes that make up a short row come into contact with the same strip of metal (Figure 7c). Wires poked into holes in different rows do not contact each other, even if the wires appear to be side by side.

In this activity you will use a breadboard to make several circuits and learn about electrical components.

1. Your teacher will show you a breadboard and explain its basic function. Once you understand how to make connections, obtain a 9-volt battery, a 1000-ohm (1 K Ω) resistor, a red LED, and 6-cm pieces of wire as needed. *Note:* The resistor is important in this circuit because it will limit the force of the current through the LED. Without the resistor, the LED would get hot and would be damaged.

2. Use the breadboard, the resistor, the LED, the battery and leads, and wires as needed to get the LED to light, similar to the way you lit the light bulb in Laboratory 1. (Hint: Always insert the LED so that the longer wire is closer to the positive terminal of the battery than the shorter wire. Later you will understand why this is necessary.)

3. Draw a circuit diagram in your notebook when you are successful in getting your LED to light.

4. Now modify your circuit by inserting a diode. Try several ways to insert it until once again you get the LED to light. Once the diode is successfully positioned, try reversing it. Record what happens. Diagram this new circuit in your notebook, noting what happened when you reversed the position of the diode.

5. Replace the resistor with a cadmium sulfide photoresistor (also called a photocell). Cover the disk part of the photocell with your fingers, and notice what happens to the light coming from the LED. Draw this new circuit in your notebook and describe what happens when you cover the photocell.

6. Finally, try reversing the way that the LED is connected to the breadboard. And notice if the LED stays lit.

Analyzing the Data

Discuss these questions with other members of your team:

- How is the circuit you designed to light the LED different from the one you made in Laboratory 1? How is it similar?
- What are the advantages in using breadboards to make electrical circuits? What are the disadvantages?
- Electronic components are used to control the flow of electric current. Explain how each of the components in the list below controlled current flow in your breadboard circuit.

  ≫diode

  ≫resistor

  ≫photoresistors

  ≫LED

Record your answers in your laboratory notebook.

Arriving at Conclusions

Discuss this question with your team members:

? *What are some of the advantages in constructing circuits with solid-state electronic components instead of larger devices such as bulbs, wires, and switches?*

Record the main points of your discussion in your notebook. note

Connecting to the Problem

In Part A, you noticed that the farther the current flowed through a material, the more resistance it encountered. Think about the chip you examined under the microscope. What is one advantage you learned in this investigation for making miniature circuits? ?

In Part B, you cooperated with your team to build a circuit containing several electrical components. Although one person can

Science Connections

How Solid-State Electronic Components Work

Many important electronic components are made from materials called *semiconductors*. Semiconductors are materials that do not normally conduct electricity well but can do so under certain conditions. Many types of materials can act as semiconductors but the most widely used by far is silicon. Silicon is the chemical element best recognized as the main ingredient in sand and glass. Pure silicon forms crystals that are shiny, brittle, and nonconductive. Inside the crystal, silicon atoms line up in a regular pattern, each sharing its electrons with the atoms near it. Silicon has four electrons available for sharing, so each silicon atom generally forms bonds with four other atoms.

At room temperature, pure silicon crystals do not conduct electricity because all of the element's electrons are occupied in making chemical bonds with surrounding atoms (Figure 8a). But, what if the crystal is not pure? In other words, what happens if the silicon is mixed with other elements?

If atoms of the element gallium are deposited on the surface of the silicon, some of the nearby silicon atoms are replaced by gallium atoms. Gallium atoms have only three electrons available for bonding. This means that the crystal can no longer form its perfect latticelike pattern made up of atoms that are all sharing four electrons. With gallium in the picture, there are some empty spaces, or holes, in the crystal structure, where electrons are missing (Figure 8b). Now it is possible for extra electrons from surrounding atoms to move into the holes. The fact that these holes attract electrons makes the crystal act like a region of positive charge. As a result, a crystal infused with an element like gallium becomes **p-type** (positive) silicon.

What happens if atoms with *more* than four available electrons are added to the silicon crystal? Arsenic atoms each have five electrons available for bonding, which means that they have four electrons for sharing with the surrounding silicon, with one extra electron per atom free to move through the crystal (Figure 9). A region where an element like arsenic is present becomes a negatively charged region of the crystal. The silicon with these extra electrons is called **n-type** (negative) silicon.

Figure 8
Pure and gallium-doped silicon.

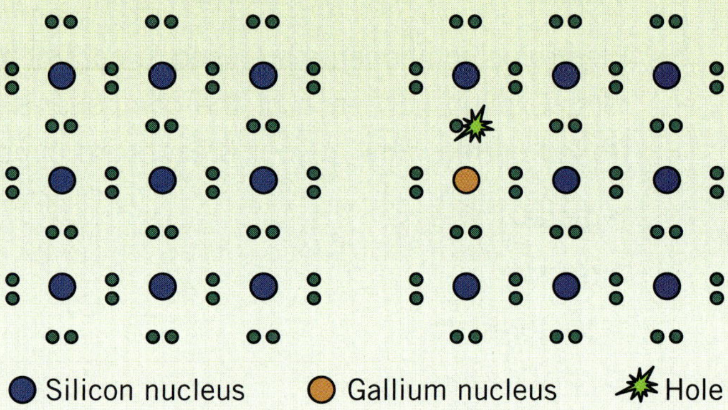

(a) A schematic drawing showing the arrangement of atoms in a sample of pure silicon

(b) A schematic drawing of a sample of silicon doped with gallium (P-type)

design, diagram, and build a circuit, there are advantages in doing it as a team. Discuss the role of your team and other teams in your class in helping build a more complex circuit.

How did this laboratory help you understand how electronic components work?

How could the investigation have been improved to help you understand more about how a circuit works?

Write your feedback about the training in your notebook. Complete a survey form provided by your teacher.

Adding atoms such as gallium or arsenic to silicon is called **doping.** Doping was the breakthrough that launched a whole revolution in electronics and integrated circuits.

So how are p-type and n-type regions organized to make an electronic component? Here is one example. To make a diode, a p-type surface is brought in contact with an n-type surface to create a p-n interface, or **pn junction.** If a battery or other voltage source is connected in one direction, the electrons are pulled across the pn junction, and the current flows to complete the circuit (Figure 10a). If the battery is hooked up in the opposite direction, the electrons are pulled away from the pn junction and electricity does not flow (Figure 10b). The result is a *diode*—an electronic component placed in a circuit to direct the flow of current in one specific direction.

This is just one example of how a semiconductor-based electronic component works. There are many additional ways to manipulate a semiconductor like silicon to form a variety of components, such as transistors, thyristors, triacs, capacitors, and relays. Some of these names are familiar; others are not. The important point about them all, however, is that they control electricity in some fashion, and each can be fabricated on silicon wafers by using various doping combinations. By creating circuits composed of these and other electronic components in a solid state on the surface of silicon chips, we can make the huge array of integrated circuits that have created the revolution in electronics.

Figure 9
A schematic illustration of silicon doped with arsenic (N-type).

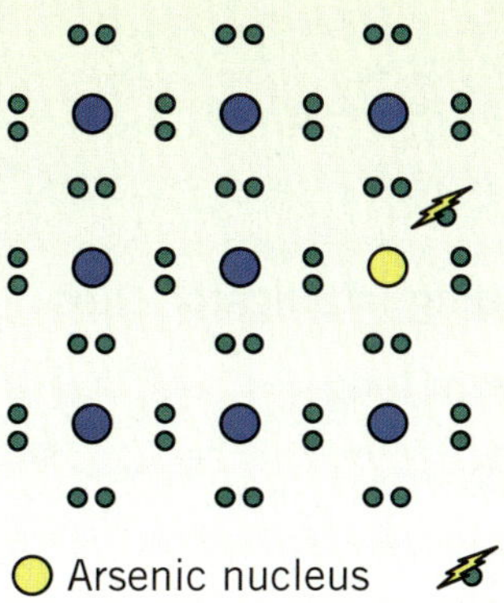

Figure 10
A diode directs the electrical current in one specific direction, while blocking its flow in the opposite direction.

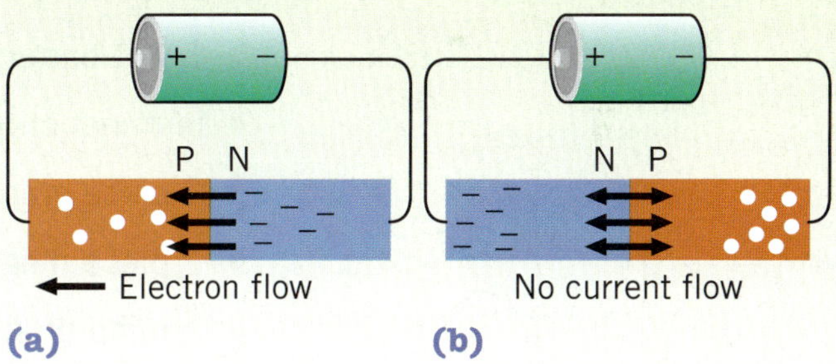

Laboratory 3 Testing Materials for Conductivity

Laboratory Introduction

In this laboratory, you will investigate the electrical conductivity of materials in order to classify them as electrical conductors or insulators.

Background

As you think back over the previous investigations, you will realize that metals play an important role in building circuits. Because electrons are free to move in metals, such materials provide an easy pathway for the flow of electrical energy. If you examine the light bulb you used as part of your circuit in Laboratory 1, you will see that the metallic pathway continues through wires and bulb filaments. In solid-state integrated circuits, some of the electronic components are specially treated regions of silicon; however, metals deliver current to and from these regions in ways similar to the wires of ordinary circuits. In building a chip, metals such as gold, silver, and aluminum may be applied to complete the circuit design. In this investigation, you will use the breadboard circuit you designed in Laboratory 2 to test a few materials to determine which of them are conductors and which are insulators.

Materials

- 1 breadboard
- 2 pieces of insulated wire (15 cm long—using the wire strippers, strip away enough insulation to expose about 0.5 cm of bare wire at each end)
- wire strippers
- 1 red LED (light-emitting diode)
- 1000-ohm (1 K Ω) resistor
- 9-volt battery
- terminal clip for the battery, with leads ending in alligator clips
- 1 piece graphite pencil lead (C), about 5 mm long
- 1 piece magnesium (Mg) ribbon, 1 cm long
- 1 lump of silicon (Si), approximately 0.5 g
- 1 small piece of tin metal (Sn), as wire or foil
- 1 small lump of sulfur (S), about 0.5 g
- 1 small piece of iron (Fe), as 1-cm steel wire
- 1 small strip of zinc or zinc (Zn) wire, about 1 cm
- assortment of small objects found in classroom, for example, pencil erasers, safety pin, glass stirring rod, notebook rings, paper, chalk, etc.

Safety

- Only use the single dry cell (9V) provided for your team.
- Use caution when handling the wire strippers. Always strip the insulation away from yourself.

- Avoid touching the metal wire ends when the circuit is complete. The wires can get hot.

Procedure (for 3 to 4 team members)

1. Assemble the breadboard circuit that you constructed in Laboratory 2, according to the diagram you made in Part B, Step 3. Include the 9V battery in the circuit and make sure that the LED lights up as before.

2. Add two 15-cm wires to the circuit in such a way that when the bare ends of the wires touch, they complete the circuit, lighting the LED. You will use this device to test the conductivity of materials.

3. Make a drawing in your notebook to record your design for a conductivity testing apparatus.

4. Make a table in your laboratory notebook similar to Table 2.

5. Use your conductivity device to test each material provided to your team. Place the bare ends of the wires close together on the material to be tested. Do not let the wires touch each other.

6. Determine if the LED is on or off as you test each object as directed in Step 5. Record your data.

Table 2

Testing Materials for Conductivity

Name of Material Tested	LED: On or Off?	Conductor or Insulator?
Carbon (graphite)		
Magnesium		
Silicon		
Tin		
Sulfur		
Iron		
Zinc		
Add rows for objects as needed		

Analyzing the Data

Discuss these questions with your teammates:

- How did you modify your breadboard circuit from Laboratory 2 to make it into a conductivity tester for this laboratory?
- What was the function of the two wires you added to the circuit? Did you need to modify the circuit in any other way?
- Which of the elements (the first seven labeled samples) were conductors of electricity?
- Which of the classroom objects were conductors of electricity?
- What do all of the conductors you identified have in common?

Record your answers in your notebook.

Arriving at Conclusions

? *Which of the elements and which of the objects could be used as electrical insulators?*

? *Examine the elements and objects that were conductors and compare their visible properties, for example, luster and texture. By what properties do the two groups differ?*

? *It is easy to explain why conductors are chosen for making circuits. Why do you think that insulators are important in designing circuits?*

Connecting to the Problem

This laboratory was designed to help you develop the concept of conductors and insulators.

? *Why is it important for a technician in the semiconductor industry to know about the conductive properties of a variety of materials?* In this laboratory, you constructed a conductivity apparatus by modifying a previous circuit design.

? *Was the diagram you made in Laboratory 2 useful in making and modifying this apparatus?*

? *What would have made your previous diagram more useful?*

In this training laboratory, you worked as a team to modify a circuit for use in a specific laboratory test. How well did your team work together to solve this problem? In a discussion with your team members, suggest ways in which your group can improve its strategy so that similar tasks in the future will go more smoothly. Make notes about the progress of your teamwork in your laboratory notebook. Fill out a teamwork survey form provided by your teacher.

Science Connections

Finding Semiconductors on the Periodic Table

There are three main groups of elements on the Periodic Table of the Elements: metals, nonmetals, and metalloids. The outer shells of metal atoms are partially filled with electrons, which are easily transferred for making chemical bonds. These easily shared electrons are also free to flow as electrical current, making metals good conductors of electricity.

Nonmetals do not conduct electricity easily since in their outer shells the electrons available for bonding do not transfer easily. As a result, nonmetals are insulators.

But metalloids lie between metals and nonmetals, both in their locations on the Periodic Table and in their characteristics. Many metalloids are semiconductors—elements that are insulators at low temperatures. As temperatures rise, some electrons begin to flow, making metalloids conductors. The prefix semi- fits them well. They are only "semi" or "partial" conductors of electricity. Silicon and germanium are metalloids. Germanium is too expensive for widespread use in the semiconductor industry. Silicon, an element found in common sand, is the second most abundant element in Earth's crust. Only oxygen is more abundant.

PERIODIC TABLE OF THE ELEMENTS

Key: 1 — Atomic number; Hydrogen — Name; H — Symbol; 1.008 — Atomic weight

Metals | Metalloids | Nonmetals

Metalloid elements: B, Si, Ge, As, Sb, Te, Po

Period	Group 1 (1A)	2 (2A)	3 (3B)	4 (4B)	5 (5B)	6 (6B)	7 (7B)	8 (8B)	9 (8B)	10 (8B)	11 (1B)	12 (2B)	13 (3A)	14 (4A)	15 (5A)	16 (6A)	17 (7A)	18 (8A)
1	1 Hydrogen **H** 1.008																	2 Helium **He** 4.003
2	3 Lithium **Li** 6.94	4 Beryllium **Be** 9.01											5 Boron **B** 10.81	6 Carbon **C** 12.01	7 Nitrogen **N** 14.01	8 Oxygen **O** 16.00	9 Fluorine **F** 19.00	10 Neon **Ne** 20.18
3	11 Sodium **Na** 22.99	12 Magnesium **Mg** 24.31											13 Aluminum **Al** 26.98	14 Silicon **Si** 28.09	15 Phosphorus **P** 30.97	16 Sulfur **S** 32.07	17 Chlorine **Cl** 35.45	18 Argon **Ar** 39.95
4	19 Potassium **K** 39.10	20 Calcium **Ca** 40.08	21 Scandium **Sc** 44.96	22 Titanium **Ti** 47.87	23 Vanadium **V** 50.94	24 Chromium **Cr** 52.00	25 Manganese **Mn** 54.94	26 Iron **Fe** 55.85	27 Cobalt **Co** 58.93	28 Nickel **Ni** 58.69	29 Copper **Cu** 63.55	30 Zinc **Zn** 65.39	31 Gallium **Ga** 69.72	32 Germanium **Ge** 72.61	33 Arsenic **As** 74.92	34 Selenium **Se** 78.96	35 Bromine **Br** 79.90	36 Krypton **Kr** 83.80
5	37 Rubidium **Rb** 85.47	38 Strontium **Sr** 87.62	39 Yttrium **Y** 88.91	40 Zirconium **Zr** 91.22	41 Niobium **Nb** 92.91	42 Molybdenum **Mo** 95.94	43 Technetium **Tc** (98)	44 Ruthenium **Ru** 101.07	45 Rhodium **Rh** 102.91	46 Palladium **Pd** 106.42	47 Silver **Ag** 107.87	48 Cadmium **Cd** 112.41	49 Indium **In** 114.82	50 Tin **Sn** 118.71	51 Antimony **Sb** 121.76	52 Tellurium **Te** 127.60	53 Iodine **I** 126.90	54 Xenon **Xe** 131.29
6	55 Cesium **Cs** 132.91	56 Barium **Ba** 137.33	71 Lutetium **Lu** 174.97	72 Hafnium **Hf** 178.49	73 Tantalum **Ta** 180.95	74 Tungsten **W** 183.84	75 Rhenium **Re** 186.21	76 Osmium **Os** 190.23	77 Iridium **Ir** 192.22	78 Platinum **Pt** 195.08	79 Gold **Au** 196.97	80 Mercury **Hg** 200.59	81 Thallium **Tl** 204.38	82 Lead **Pb** 207.2	83 Bismuth **Bi** 208.98	84 Polonium **Po** (209)	85 Astatine **At** (210)	86 Radon **Rn** (222)
7	87 Francium **Fr** (223)	88 Radium **Ra** (226)	103 Lawrencium **Lr** (262)	104 Rutherfordium **Rf** (261)	105 Dubnium **Db** (262)	106 Seaborgium **Sg** (263)	107 Bohrium **Bh** (264)	108 Hassium **Hs** (265)	109 Meitnerium **Mt** (268)	110 Ununnilium **Uun** (269)	111 Unununium **Uuu** (272)	112 Ununbium **Uub** (277)		114 Ununquadium **Uuq** (285)		116 Ununhexium **Uuh** (289)		118 Ununoctium **Uuo** (293)

Lanthanides	57 Lanthanum **La** 138.91	58 Cerium **Ce** 140.12	59 Praseodymium **Pr** 140.91	60 Neodymium **Nd** 144.24	61 Promethium **Pm** (145)	62 Samarium **Sm** 150.36	63 Europium **Eu** 151.96	64 Gadolinium **Gd** 157.25	65 Terbium **Tb** 158.93	66 Dysprosium **Dy** 162.50	67 Holmium **Ho** 164.93	68 Erbium **Er** 167.26	69 Thulium **Tm** 168.93	70 Ytterbium **Yb** 173.04
Actinides	89 Actinium **Ac** (227)	90 Thorium **Th** 232.04	91 Protactinium **Pa** 231.04	92 Uranium **U** 238.03	93 Neptunium **Np** (237)	94 Plutonium **Pu** (244)	95 Americium **Am** (243)	96 Curium **Cm** (247)	97 Berkelium **Bk** (247)	98 Californium **Cf** (251)	99 Einsteinium **Es** (252)	100 Fermium **Fm** (257)	101 Mendelevium **Md** (258)	102 Nobelium **No** (259)

How a Transistor Works

The most important components of any integrated circuit are *transistors,* devices that allow tiny amounts of electrical current in one circuit to control the flow of much larger amounts of electrical current in another. A typical transistor embedded in a silicon chip is built like a semiconductor sandwich. A p-type region in the middle, called the *gate,* is surrounded by two n-type regions; one is called the *source,* and the other, the *drain.*

With a surplus of available electrons on both sides, the electron-lacking holes in the p-type gate are soon filled, leaving this as an uncharged, nonconducting zone—an electronic logjam. When current is applied at the source, there is no conducting pathway through this charge-depleted gate. In this state, the transistor acts like any switch in the "off" position. It simply breaks the circuit.

But apply even the smallest amount of positive charge to the gate, and this condition changes rapidly. Once again there are electrons flowing freely in the central gate region, forming a conductive pathway through which current flows from the source to the drain. In this state, the transistor acts like a switch in the "on" position, completing the circuit.

On a silicon chip, a typical transistor is integrated into the chip. In Figure 11, a positive gate voltage attracts electrons into the underlying p-type silicon, creating a thin, n-type channel between the source and drain through which current can flow.

With no moving parts or delicate components to wear out, transistors are very reliable for use in integrated circuits. Even better, the fact that they use so little energy means they generate almost no heat as they operate, allowing many of them to be packed into a small area. With transistors becoming smaller all the time, the speed at which they operate is increasing accordingly. Researchers are experimenting with transistors made out of carbon "nanotubes" and single molecules, predicting a future with even faster, more complex electronic devices for an ever-increasing number of applications.

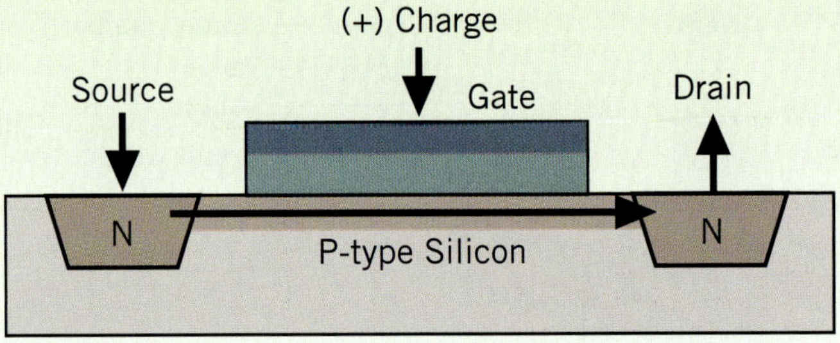

Figure 11
Cross-section of a transistor.

Laboratory 4 Adding a Transistor to a Breadboard Circuit

Laboratory Introduction

In the Science Connection, you saw how transistors can be integrated into the circuitry of a chip by combining p and n regions in an appropriate pattern. In this laboratory, you will observe how transistors work by adding a solid-state transistor component to the breadboard circuit you made in Laboratory 2.

Background

Examine the small solid-state transistor provided with your materials for this laboratory. Other components you have used—LEDs, diodes, and resistors—have had two wire leads to insert into the breadboard. The transistor has three. One represents the source, one the drain, and one the gate. Be sure that you can identify each one before you begin your work. Think about how you will place a transistor in your circuit. You will want current to flow *through* the transistor *from* the source *to* the drain. The lead for the gate supplies the positive charge necessary for the current to flow through the transistor. The source of that positive charge comes from a second circuit you will build.

Materials

- breadboard
- several pieces of insulated wire (6 cm long—using the wire strippers, strip away enough insulation to expose about 0.5 cm of bare wire at each end)
- wire strippers
- red LED
- 1000-ohm (1 K Ω) resistor
- 9-V battery
- terminal clip with leads for battery
- 1.5-V battery
- battery holder with leads for 1.5-V battery
- 2 15-cm pieces of insulated wire (using the wire strippers, strip away enough insulation to expose about 0.5 cm of bare wire at each end)
- 1 n-channel transistor for breadboard circuit

Safety

- Have your teacher check your circuit designs before attaching the power sources.
- Use care when removing the insulation from the ends of the wires.
- Avoid touching bare wires or leads since they may get hot.

Procedure (for 3 to 4 team members)

1. Referring to the basic circuit diagram you drew in Laboratory 2, Part B, reconstruct the breadboard circuit that includes the resistor and the LED. After your teacher checks your construction, attach the 9-V battery as you did in Laboratory 2 and test it to confirm that the LED lights up.

2. Disconnect one battery lead from the basic circuit. Use Figure 12 to devise a way to add the transistor to the circuit, rearranging the positions of the other components as necessary. The transistor *source* should be closest to the (+) terminal of the battery, and the *drain*, closest to the (–) terminal of the battery. The *gate* should not be included in the basic circuit at this point. Do not reattach the battery. Draw your circuit design, and make a prediction: Will the LED light up when the battery is reattached? Record and explain your prediction in your notebook.

3. Have your teacher approve your design before reattaching the battery. Was your prediction correct? Did the LED light up? Record your result. If your prediction was not correct, make some corrections in the explanation that you recorded in number 2.

4. Place the 1.5-V battery in a battery holder with leads. Design a second circuit for supplying voltage to the gate of the transistor. Figure 12 is a basic guide for this addition. You will need to translate it into breadboard circuitry.

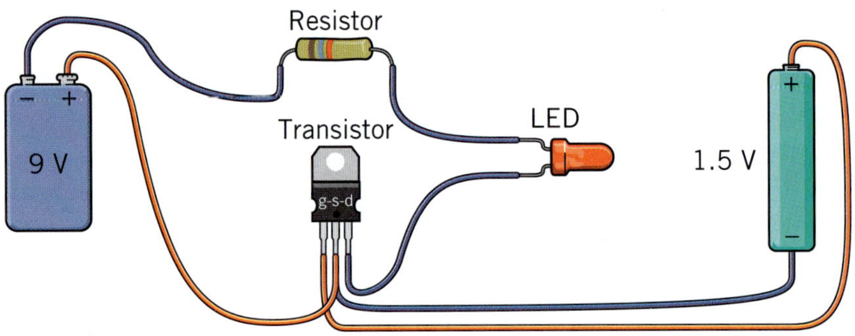

Figure 12
A circuit that includes a transistor.

5. Arrange the components—except for the two batteries—on your breadboard so that the circuit pattern follows the diagram. Get your teacher's approval before including the batteries. Make a prediction. Will the LED be illuminated when you attach both batteries? Record your prediction and explain your reasoning in your notebook before you go to the next step.

6. Attach the batteries and note whether or not the LED lights. Record the result in your notebook. Was your prediction correct? Why or why not?

7. Break the circuit that includes the 1.5-V battery by detaching one of the wires from the breadboard. Note the result in your notebook.

Analyzing the Data

- Did adding a transistor to the basic circuit break or maintain the current flow?

- How were you able to modify the transistor so that it permitted current to flow?
- On what evidence do you base your answers to these questions?

Arriving at Conclusions

In this circuit investigation, the transistor first acted like a switch in the "off" position and later acted like a switch in the "on" position. Go back to the Science Connection on transistors on page 32 and explain how you were able to make the transistor that you added to your circuit perform as an electrical switch.

What was the function of the second electrical circuit in the operation of your transistor?

Connecting to the Problem

The purpose of this laboratory was to increase your understanding of the way transistors operate in an electrical circuit. Recall that thousands of transistors are often integrated into a single silicon chip.

How do you think transistors contribute to the flow of electricity through the maze of integrated circuits on a single chip?

What advantages are there in developing and integrating smaller and smaller transistors into chips?

Your training personnel are interested in monitoring how well your training is advancing your knowledge of

semiconductors

integrated circuits

transistors

The facility is also interested in how well you are completing tasks as a member of a workplace team. Review the goals of this training module on page 21 and complete a survey provided by your teacher.

Laboratory 5 Making an Integrated Circuit on a Chip: A Model

Laboratory Introduction

You read in the Industry Overview on pages 3–6 that manufacturing integrated circuits on semiconductor wafers involves a series of masking and etching steps in which layer after layer of doping atoms and metal leads are added to the silicon.

In this training laboratory, you will model some of these steps in order to visualize how all of the specialized manufacturing procedures are related to the end product—the semiconductor integrated circuit.

To prepare for these modeling steps, review the Industry Overview, the video, and the Overview section on the CD-ROM. In particular, carefully note the sequence of steps involved in masking, doping, and photo-developing the wafer as it is fabricated.

Part A Modeling Fabrication: Masking and Doping

Background

An integrated circuit is constructed in a series of layers, each of which has different electrical properties. When the layers are completed, different regions become transistors, diodes, resistors, and many other electrical components. Together with tiny metal wires, the layers form the integrated circuit.

Materials

- copies of 3 masks provided by your teacher, each on a piece of 8 1/2 × 11-inch card stock
- craft knife or small scissors with sharp points
- sheet of heavy cardboard to protect desktop
- 2 colored pencils, one red, one blue
- #2 pencil
- 1 plain piece of copier paper, 8 1/2 × 11-inch

Safety

- Be very careful in handling the craft knife or small scissors. Always cut away from yourself and others, keeping your other hand away from the blade.
- Return blades and scissors to their protective packages, and return them as directed by your teacher.

Procedure (per pair of students on teams)

1. Obtain the first sheet of card stock, marked "P." Lay the card stock on the corrugated cardboard, and carefully cut inside the lines until you have cut out all of the shapes in the card stock.

2. See Figures 13 and 14. Lay the P mask on the plain piece of copier paper, carefully aligning the edges. Using the P mask as a

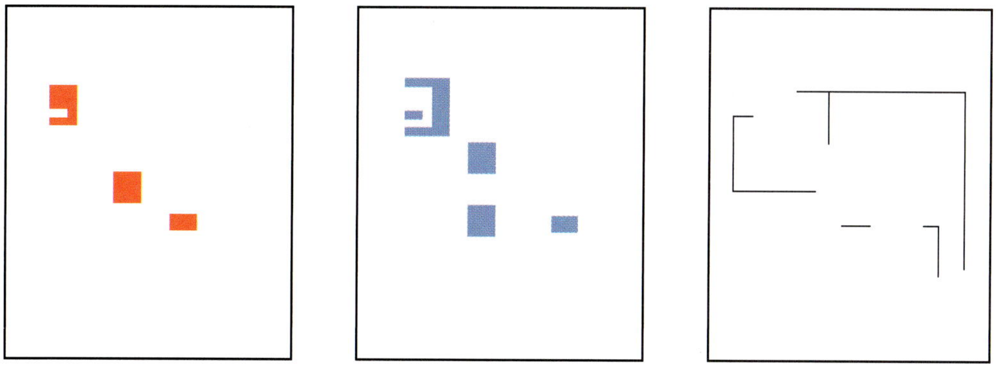

(a) The P mask (b) The N mask (c) The wire mask (M)

Figure 13 Masks.

Figure 14 Coloring in the P mask.

stencil, use your red pencil to color the shapes onto the piece of plain copier paper. Remove the P mask.

3. Obtain the second sheet of card stock, marked "N." Repeat the procedure in Step 1 for cutting out the shapes.

4. Lay the N mask on the same piece of copier paper, and, with a blue pencil, repeat the procedure in Step 2 for transferring the shapes to the copier paper.

5. Obtain the cardstock marked "M." Repeat the procedure in Step 1 for cutting out the shapes.

6. Lay the M mask on the copier paper. Use the #2 pencil to transfer the shapes to the copier paper.

Analyzing the Data

Examine your completed diagram on the copier paper. You have created a model of a circuit that is created in three layers. Red shapes represent P-type silicon, blue shapes represent N-type silicon, and pencil lines represent metal wires.

- Find a diode on your model circuit. Label it, and explain why it is a diode. *note*

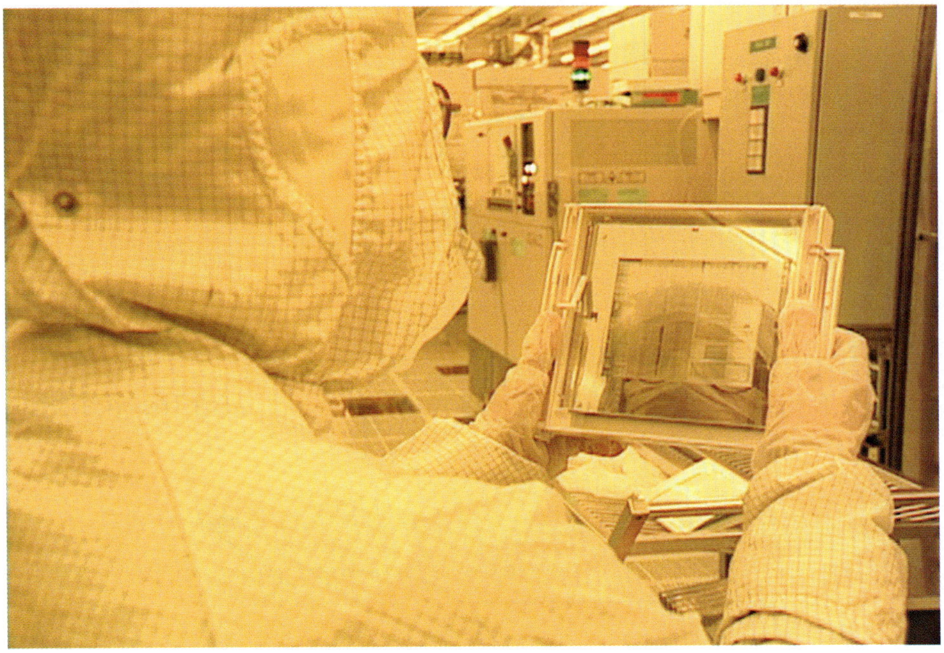

A technician checks a glass lithographic plate onto which masks will be projected.

- Find a transistor on your model circuit. Label it, and explain why it is a transistor.
- Where would current enter the circuit you modeled? Where would it exit?

Arriving at Conclusions

Use the information in the Industry Overview on pages 3–6 to describe the steps in the chip manufacturing process that you modeled in Part A.

Part B Photolithography

Background

As each layer of the chip is formed, the technician uses great care to align the masks perfectly, one on top of the other. To make this positioning even more challenging, each mask is projected onto a surface hundreds of times smaller than its original size.

In this modeling activity, you will simulate the way that masks are positioned and scaled onto chips. The *Making Semiconductors* video illustrates how the process occurs in a production plant.

Materials

- set of 3 masks provided by your teacher, each on a sheet of ordinary copier paper
- completed model from Part A
- access to a copy machine capable of reducing an image to at least 70%

- ruler
- calculator

Safety

Use the copy machine under your teacher's supervision. Never look directly at the moving light.

Procedure (for 3 to 4 team members)

1. Your teacher will provide a fresh copy of the model you used from Part A. Draw a 15-cm × 15-cm square that encloses the completed model from Part A. Just below the square you drew, draw a second square 7 mm × 7 mm to represent the size of a typical chip. The object of this modeling activity is to transfer the layers of the integrated circuit, one mask at a time, to fit within an area 7 mm × 7 mm.

2. Find out the minimum percent size reduction you can use to photoreduce a figure with your copy machine.

3. Calculate the number of times you will need to reduce a mask in order to fit it onto a square that is approximately 7 mm × 7 mm. *Hint:* Suppose that your copy machine is capable of reducing a figure to 70% of its original size. Beginning with a figure that fits in a 150 mm × 150 mm square, you would reduce it to one that fits into a 105 mm × 105 mm square on the first copy attempt (150 mm .70 = 105 mm). Continue to reduce the product by 70% in a series of steps until the final product is equal to or smaller than 7 mm × 7 mm.

4. Beginning with a copy of the P mask diagram, make the calculated number of copies. Be careful to place the sheet of paper in exactly the same position in the paper tray each time you make a new reduction. Keep all of the copies in the order of their reduction.

5. Repeat the same procedure using the N mask. With careful planning, you can use your sequence of P mask reductions in the paper tray. Keep these P-N combinations in the order of their reduction.

6. Finally, repeat the same procedure using the M mask (the one for the wires in the circuit). Again, with careful planning, you can use the P-N reductions in the paper tray.

Analyzing the Data

- How does your final product compare to the image of the complete circuit you modeled in Part A?
- How many times did you calculate that you would need to photo reduce the images in order to achieve the dimensions of a chip? Show your calculations in your notebook. Were you correct? Explain.

- How successful were you in getting the masks to superimpose on one another to assemble the same circuit that you modeled in Part A? If you were unsuccessful in superimposing the images correctly, how would you change your procedure in another trial?

Arriving at Conclusions

In this modeling activity, a photocopier served to develop images of circuitry components on paper. Review the Industry Overview on pages 3–6. How does this photocopying procedure model the steps of making an integrated circuit on a chip? ? note

Connecting to the Problem

In the semiconductor industry, technicians perform a lengthy series of specialized tasks to transfer a circuit design to a silicon chip. In your training, you are learning about the "big picture" or the entire process in order to understand how each step contributes to the final product.

How effective are these modeling activities in building your understanding of the manufacturing process? How could they be improved? Write your comments in your notebook.

You may have noticed some manufacturing steps that were not directly modeled in this part of the training. For example, how would you suggest modeling the etching process that takes place after a mask is applied? What are some other steps that might be modeled effectively? Discuss some ideas with your team, and record them in your notebook. note

? *How well did your team work together in performing these modeling activities? Record your assessment of your progress in working together, and fill out a survey provided by your teacher.* note

Laboratory 6 Evaluating the Cleanliness of a Workplace

Laboratory Introduction

In this investigation, you will invent a method for analyzing the cleanliness of the air in your workplace (your classroom). You will compare your air quality to that of the air in a semiconductor fabrication "clean room." By doing so, you will understand the meaning of "clean" in a semiconductor fabrication work area.

Background

Most of the fabrication process for turning semiconductor silicon wafers into integrated circuits takes place in what the industry calls a "clean room." Clean, in this context, is an understatement! Semiconductor industry clean rooms are said to be 10,000 times cleaner than the operating rooms of hospitals.

The meaning of "clean" includes the people working in a clean room. The coverall "bunny suits" you see technicians wearing in the video take a long time to put on but are essential in maintaining a clean manufacturing environment.

Materials

- glass slides
- coverslips
- slide-cleaning paper
- hand lenses
- compound light microscope
- clear tape
- plastic deli container with lid (very clean)
- zip-closing plastic bags
- stereomicroscope

Multimedia Links

VCR CD-ROM Review the video as well as the CD-ROM to find a magnified image showing the effect of a speck of dust on the circuitry of a chip.

Safety

- Be careful when handling glass slides and coverslips.

Procedure (for 3 to 4 team members)

1. Use some or all of the available materials to invent a way to count the number of dust particles that settle in any one area of your classroom over a 24-hour period.
2. When you have agreed on a plan, review it to make sure it includes a
   - control
   - method of defining space
   - method of counting dust particles
   - method of verifying your results
3. Record your plan and procedure in your notebook.

Science Connections

Water and Air—Clean to the Extreme

Water is used in many steps in the semiconductor fabrication process. Since any impurities may damage the ultrafine circuitry of a computer chip during manufacturing, the water used is cleaned of any and all contaminants. That is why semiconductor fabrication facilities maintain their own highly specialized water treatment equipment. Water is passed through gigantic columns of activated carbon to remove some impurities (Figure 15).

Another step filters the water by means of a process called reverse osmosis. In osmosis, particles diffuse, or move, from the side of a membrane where they are most prevalent to the side where they are less prevalent—they are said to move along a gradient from the area of higher concentration to area of lower concentration. Reverse osmosis uses pressure so that the water moves through a membrane against the gradient—from an area where there are more impurities to a place where there are none. The special membrane removes very tiny particles (even those the size of ions) so the water that results is pure.

The air in the "clean rooms" of semiconductor fabrication facilities must be filtered to meet a standard of no more than one particle of dust for every cubic foot of air. Fabs achieve this ultraclean air by drawing the air down through the floor and completely filtering it on an average of 10 times per minute. Workers in the clean room must wear special lint-free, anti-static "bunny suits" to avoid leaving particle impurities in the air or on any surfaces. One tiny particle of dust can contaminate an entire wafer, rendering it useless. Since the success of a semiconductor fabrication facility depends on the percent of products that are unflawed (percent yield), any flawed wafers hurt the company's productivity rating.

Figure 15
Activated carbon filters in a semiconductor fabrication facility.

4. Your teacher will suggest a way to distribute sampling areas throughout the room. This will enable you to compare different areas of the room.

5. Show your plan to your teacher for approval.

6. Conduct your sampling and collect your data. Repeat the sampling and data collection over a second 24-hour period.

Science Connections
Robots

Robots are machines that can be programmed to perform tasks that would otherwise have to be done by humans. In industry, robots are particularly useful in performing repeated tasks that require great precision in an ultraclean environment. In other settings, robots can be designed to perform tasks too risky for humans, such as collecting samples of rocks from other planets, fixing parts of spacecraft, retrieving objects from the ocean floor, collecting data in volcano craters, or welding at very high temperatures.

Modern robots contain microprocessors, chips with circuits capable of organizing and delivering complex information. On these chips, programmers write coded instructions that direct the robots to respond to stimuli, perform very precise tasks, repeat tasks in exactly the same way each time, and carry out many other functions.

In the semiconductor industry, robots are programmed for such tasks as sorting and inspecting wafers. The sort consists of an electrical test on each chip to determine whether or not the chip meets specifications. If it does not, it gets marked so that it doesn't end up getting packaged with other chips that do meet specifications. Since the yield of "good" chips determines the success of the fabrication facility, robots play an important role in maintaining the industry's productivity.

A robotic arm moves wafers inside an ultra-clean machine.

Analyzing the Data

Review your particle data for the two 24-hour periods.

How do the data for these periods compare? Is it meaningful to average the counts over the two days? Why or why not?

Compare your data to those of other groups in other parts of the classroom. Where was the most dust collected? How could you account for that?

Arriving at Conclusions

Compare the cleanliness of your classroom to that found in a semiconductor fabrication clean room (one particle per cubic foot). Use the data you collected for your own workspace, and mathematics to describe how much cleaner a clean room is than your classroom. Compare your calculations to those of other teams. Record your calculations and your findings in your notebook.

Connecting to the Problem

Sometimes technicians are required to follow established protocols for accomplishing a task, and sometimes they must devise their own methods for solving workplace problems. In this training step, your team worked together to invent a method for collecting data on airborne particles.

? *How well did your team work together to exchange ideas before making a decision?*

? *How can the teamwork be improved?*

Record your evaluation of the way your team accomplished this step of the technician training program. Fill out a survey provided by your teacher.

Part of your responsibility as a technician in the semiconductor industry is to maintain the mandatory cleanliness of the work area. In your notebook, explain how this responsibility qualifies as a team effort. note

Making and Testing a Robot

Overview of the Task

Throughout your semiconductor industry technician training program, your team has worked together to gather background information, interpret and follow protocols, wire circuits, test components, test the conductivity of materials, model chip manufacturing, and analyze air purity. Now your team will receive the materials, solid-state components, and directions needed to build a robot.

Performance Evaluation Criteria

The training staff will monitor your team's progress in completing this final task. Your performance will be judged on how well you

- follow the protocols needed to wire the circuitry
- identify and describe the function of the solid-state components included
- discuss the overall design of the robot
- devise a test for the robot's performance
- suggest modifications that would alter the robot's behavior

The Rocky 4 Pathfinder was the prototype for the Mars Rover. The Rover was delivered to Mars in the Sojourner mission and traveled over Mars' surface collecting soil and rock data.

- suggest several uses for the constructed robot, both in industry and in the consumer market
- evaluate your team's success in working together on this task

Materials

- BYO-bot robot kit
- pair of sharp scissors or a craft knife
- forceps with sharp points or a needle-nosed pliers
- flashlight with strong beam
- 1 photo resistor from Laboratory 2
- 1 transistor from Laboratory 4
- 1 LED from Laboratory 4

Safety

- Be careful when handling scissors or knife and pointed forceps or pliers. Direct sharp edges and points away from you when handling these tools.
- Do not touch bare wires during operation of the robot, since they may get hot.

Part A Examining the Solid State Components

Included in the kit for building your robot are two transistors and two light sensors. As you examine these components, make notes in

Robot-welders at work on an automobile assembly line.

your laboratory notebook. Later you may be asked questions about these components and their uses to assess your understanding of solid-state components.

Transistors

- Examine the two transistors in this kit and compare their appearances to the one you used in Laboratory 4. Which features are the same? Which are different?
- When the kit transistor is inserted into the robot circuitry, the transistor will perform similarly to the transistor used in Laboratory 4. What conditions will be needed for it to act like a switch?
- Later, when you follow directions for inserting the transistor into the circuit, you will need to be careful about inserting the three wires in exactly the correct order. Why will this be important for the transistor to function?

Record your observations and explain your reasoning in your notebook.

Light Sensors

Examine the light sensors in the kit. Recall your use of another light sensitive solid-state component, the photo resistor, in Laboratory 2. Compare the appearance of the photo resistor to these light sensors. How are they the same? Different? Make sketches and notes in your notebook to compare the two components.

- Although you will use the light sensors to control the current in the robot circuit, they are actually more closely related to another solid-state component you used in Laboratory 2. Which one? *Hint:* examine the wires leading from the sensor. Notice that one is longer than the other.

The light sensor is a type of diode. But the one in this kit acts in a way that is opposite to the LED. It does not emit light when it is placed in an electrical circuit. Instead, when placed in the light, it generates free electrons in the p-region. That means that when this solid-state sensor is wired into a circuit, current will flow in the circuit only when the sensor is placed in light. And, within a certain range, the more light to which the sensor is exposed, the more current will flow.

- Later, when you follow directions for inserting the light sensor into the circuit, be careful about inserting the two wires in exactly the correct order. Why will this be necessary for the light sensor to function?

Record your observations and explain your reasoning in your notebook.

Transistors and Light Sensors in a Circuit

You will build your robot by wiring together dry cells, motors, transistors, and light sensor diodes. For now, think of the motor as a user of electricity, much like the bulb you connected in Laboratory 1.

- In your laboratory notebook sketch a BYO-bot design that includes the battery, the motor, the transistor, and the light sensor. Show where the wires will be attached. Make the diagram very simple, using the one in Laboratory 4, page 32, as a guide. As in that laboratory, use electrical symbols.

Label your diagram with the names of the components, and write a brief explanation of how the BYO-bot operates.

Part B Building the Robot

- You will need the BYO-bot kit, a cutting tool, and a forceps.
- Work together as a team to follow the assembly directions included in the kit. Be prepared to describe organization strategies your team has adopted during the course of your technician training. How are decisions made? How are roles assigned?

Be prepared to explain each step of the assembly to other members of your own team or to the training staff monitoring the task.

- Follow safety instructions when using sharp tools in the assembly. Be prepared to discuss the importance of safety in the semiconductor workplace.

Part C Testing the Performance of the Robot

- Obtain a flashlight. Find a large, smooth surface, preferably the floor, to test your robot.
- Shine light on the robot to test its performance. If it does not respond to direct light, go back and check the insertions of components and wires.

Devise several tests in which you vary the light stimulus. Record each test variable and way that the robot responds. *Hint:* Try varying the distance of the light source, the intensity of the light, or the orientation of the light beam. In your notebook, offer an explanation for each response.

Devise several tests in which you vary the position of the light sensors. For example, point them both up, both to the sides, or one up and one down. Be very careful when bending the sensors. It may work best to remove them, reshape them, and reinsert them into the robot. For these tests, keep the light source in the same position. Why? In your notebook, offer an explanation for each response.

- Your teacher or training staff members may request each group to demonstrate and explain these performance tests to the entire class. If asked, your group should be prepared to conduct a short demonstration and explanation of the test variables and results.

Part D Suggesting Industry and Consumer Applications

Your robot is a model for the many kinds of robots used in industry and in the consumer market.

As a team, discuss possible industrial applications for light-sensitive robots. Make a list of all the possible uses of such robots in an industry like the semiconductor industry. Extend your thinking to other industries with which you are familiar. For each suggestion, describe the task the robot might perform and any design modifications that you would need to make. Be prepared to share your list with the entire class.

As a team, discuss possible consumer applications for light-sensitive robots. Make a list of all the possible products that might appeal to consumers. Some products might be entertaining, others useful for performing household tasks. For each suggestion, describe how the light-sensitive robot would perform and how you would need to modify your design to make the product. Be prepared to share your list with the entire class.

Part E Reflecting on the Team-Building Goal of Your Training

Reflect on the work you have done as a participant in the training process for Vertical. Look back through notes you made about your team performance at the end of every training step. Discuss these questions with your team, and with other teams in your class:

- How did your laboratory work help you understand the science and technology of integrated circuits and semiconductors? What else could have been included in the module to help you understand more?
- What is the importance of teamwork in the wafer fabrication process? What happens when a team or team member does not perform as expected?

How did you like working as a technician in the semiconductor fabrication industry? Is this a job that you might want to consider as a career? How do you like solving problems and working with instruments?

Respond to these questions by writing a report to the training staff of Vertical Semiconductors. Recall that the a goal of the industry is to create a work environment based on strong teamwork and a basic understanding of all aspects of the manufacturing process on the part of all technicians. To help you prepare the report, your teacher will distribute copies of survey results collected over the course of the training. Does the collected data show that the training served to build teamwork? Refer to the survey data to explain your answer. To make your explanation clear, you may want to make a data table and a graph. Finally, remember that suggestions for improving the training are important to the company. Include them in your report. note

glossary

Activated carbon: Also called activated charcoal; powdered or grainy material (carbon) that cleans and purifies by absorbing particles.

Breadboard: Plastic plate perforated with rows of holes for holding and connecting wires and electrical components, used to build temporary circuits.

Bunny suit: Clean protective clothing worn by a semiconductor technician.

Chip: An integrated circuit; a small wafer of semiconductor material that is the basis for an integrated circuit.

Circuit: See electrical circuit.

Circuit schematic: Also called a circuit diagram; a design for an integrated circuit that is translated into a series of photomasks.

Clean room: A room for fabricating silicon chips in which an air quality standard of one particle of dust per cubic foot is maintained by constant filtering.

Conductivity: A substance's ability to conduct or transmit electrical current.

Conductor: A substance that allows electrons to flow.

Crystalline lattice: A stacking arrangement of atoms in a regular, repeating pattern.

Crystal structure: A substance's particular crystalline lattice, or stacking pattern.

Diode: An electronic component placed in a circuit to direct the flow of current in one direction.

Drain: In a transistor, an n-type (negative) region toward which charge carriers move.

Doping: Embedding atoms in the crystalline lattice of silicon to create zones of electrical conductivity.

Dry cell: A battery.

Electrical circuit: Pathway along which millions of moving electrons stream to form an electrical current.

Electrical component: A device that controls the direction and amount of electron flow.

Electrical current: Moving (flowing) electrons.

Etching: Producing a design on a hard material by eating away at its surface with a chemical or radiation.

Gate: In a transistor, a p-type region that controls current flowing through the transistor according to applied voltage.

Insulator: A substance that does not allow electrons to flow.

Integrated circuit: A complete network of electrical components fabricated on a thin silicon chip to fulfill desired applications.

LED (light-emitting diode): A semiconductor diode that gives off light in response to voltage.

Metal: A material possessing properties such as luster, ductility, conductivity, and malleability.

Metalloid: A material with properties intermediate between those of metals and nonmetals.

Microprocessor: A chip (an integrated circuit) with circuits capable of organizing and delivering complex information.

Multimeter: A tool that can make several kinds of electrical measurements.

Nonmetal: A material possessing properties such as brittleness, lack of luster, and nonconductivity (it acts as an insulator).

N-type: Relating to the negatively charged substance in a semiconductor in which charge is carried by electrons.

Ohm: The unit for electrical resistance; one ohm equals the resistance of a circuit in which an electrical potential difference of one volt produces a current of one ampere.

Percent yield: Ratio of unflawed to flawed wafers produced by a company.

Photolithography: A process used in making integrated circuits that involves the photographic transfer of a pattern to a surface for etching.

Photomask: Also called a mask; a material used to translate the pattern of a circuit schematic onto a silicon wafer; the mask material shields selected areas of the wafer's surface during the etching process.

Photoresist: A light-sensitive substance that loses its resistance to chemical etching when exposed to radiation; used to treat a silicon wafer to ensure that the pattern of a photomask transfers exactly to the wafer.

Photoresistor: Also called a photocell or photoelectric cell; an electronic device whose properties change in response to light.

PN junction: The interface between a p-type (positive) surface and an n-type (negative) surface in an electronic component such as a transistor.

Probe: A metal point used to check part of an electrical circuit.

Protocol: A detailed plan for a procedure.

P-type: Relating to a positively charged substance in a semiconductor in which charge is carried by holes.

Resistance: A property of a material based on the tendency of particles to interfere with electron flow.

Reverse osmosis: A process that uses pressure against a gradient to move very tiny particles out of a sample of water in a direction opposite to that of particle diffusion.

Robot: A machine that can be programmed to perform tasks that would otherwise have to be done by humans.

Semiconductor: A material that does not conduct electrical current under normal conditions, unless its conducting properties are altered by embedding atoms of another element in its crystalline lattice. See *doping*.

Silicon: A nonmetallic chemical element, widely used as a semiconductor, that is the main ingredient in sand and glass.

Silicon chips: Small, rectangular segments of silicon wafers fabricated to include integrated circuits.

Solid state: used to describe electronic components and integrated circuits; in physics, relating to the arrangement of ions, molecules, subatomic particles, and holes in the crystal structure of a solid substance; solid-state components incorporate semiconductor devices (rather than vacuum tubes) to control the flow of electrons.

Solvent: The dissolving agent in a solution, usually the component present in the larger quantity.

Sort: A task, performed by robots in the semiconductor industry, consisting of an electrical test on each chip to determine whether or not the chip meets specifications.

Source: In a transistor, an n-type (negative) region that supplies charge carriers for electron flow.

Switch: A device used to make, break, or change connections in an electrical circuit.

Team ethic: Training method based on the observation that technicians who were trained in more than one task and who understood clearly how each task contributes to making a product that meets specifications have a greater sense of ownership of the product; a greater sense of ownership leads to greater employee job satisfaction and increased product yield for the company.

Transistor: A device for allowing tiny amounts of electrical current in one circuit to control the flow of much larger amounts of electrical current in another circuit.

UV (ultraviolet) light: Electromagnetic radiation with a wavelength shorter than that of visible light and longer than that of X-rays.

Vacuum tube: A device for controlling the flow of electrons in which the gas inside of a glass or metal tube is removed to the level of a vacuum.

Volt-ohm meter: See Multimeter.

Wafer: A thin slice of a semiconductor material used as a base for an electrical circuit or component.

additional resources

Fabless Semiconductor Association
http://www.fsa.org
Simulates technology and foundry capacity by communicating the future needs of the fabless semiconductor segment and provides interactive forums for the benefit of its members.
Three Lincoln Center
5430 LBJ Freeway, Suite 280
Dallas, TX 75240
Telephone: (972) 866-7579

Semiconductor Research Corporation
www.src.org
P.O. Box 12053
Research Triangle Park, NC 27709-2053
Telephone: (919) 941-9400
A bit advanced but worth exploring. Plans and directs a long-term applied research program in semiconductor technologies.

Semiconductor Assembly Council
www.sacouncil.org
1901 Old Middlefield Way, Suite 15
Mountain View, CA 94043
Telephone: (650) 390-8575
SAC provides a forum for the exchange of best practices and technical information related to semiconductor manufacturing.

credits

Cover photos:	chip: David C./CORBIS; pills: Jean-Pierre Lescourret/CORBIS; vials: Tek Image/Science Photo Library/Photo Researchers, Inc.; police tape: PhotoDisc, Inc.; cows: PhotoDisc, Inc.; pipeline: SuperStock, Inc.
p. i (bottom and repeated throughout):	Kelly Harriger/CORBIS.
p. vi:	Peter Menzel/Stock, Boston.
p. ix (top and repeated throughout):	Kelly Harriger/CORBIS.
p. 3 (bottom):	David Shopper/Stock, Boston/ PictureQuest.
p. 14 (left):	Charles O'Rear/CORBIS.
p. 14 (right):	Charles O'Rear/CORBIS.
p. 45:	James A. Sugar/CORBIS.
p. 46:	IFA Bilderteam/eStock Photography/ PictureQuest.
All other photos:	American Chemical Society/Take One Video.